CONSTRUC

FINANCING
INFRASTRUCTURE PROJECTS

ONE WEEK LOAN

CONSTRUCTION MANAGEMENT SERIES

FINANCING INFRASTRUCTURE PROJECTS

TONY MERNA and CYRUS NJIRU

Published by Thomas Telford Publishing, Thomas Telford Ltd, 1 Heron Quay, London E14 4JD
URL: http://www.thomastelford.com

Distributors for Thomas Telford books are
USA: ASCE Press, 1801 Alexander Bell Drive, Reston, VA 20191-4400
Japan: Maruzen Co. Ltd, Book Department, 3–10 Nihonbashi 2-chome, Chuo-ku, Tokyo 103
Australia: DA Books and Journals, 648 Whitehorse Road, Mitcham 3132, Victoria

First published 2002

Further titles in the series

Appraisal, risk management
Engineering management
Procurement and construction management
Construction management
Operation and maintenance

A catalogue record for this book is available from the British Library
ISBN: 0 7277 3040 1

© T. Merna and C. Njiru and Thomas Telford Limited 2002

Foreword

Increasing pressure on government budgets worldwide has meant that public funds are no longer sufficient to finance all needed infrastructure projects. The growing demand for higher levels of infrastructure services has refocused attention on the use of private finance for public projects, rather than the traditional methods of public funding. This has resulted in a considerable number of infrastructure projects being funded using project finance in both developed and developing countries around the world.

Financing Infrastructure Projects is designed as a handbook, which brings together international experts to present the latest thinking and developments in the effective implementation of finance for projects. This book provides an insight into the practicability of financing different types of projects and the commercial aspects affecting projects.

The authors have for many years been actively involved in research and also in the practical implementation and management of infrastructure projects financed through different financial packages. With the changing global trends on financing of projects, more emphasis has lately been on infrastructure projects procured through public–private partnerships.

Tony Merna and Cyrus Njiru

Preface

Public finance is the traditional source of funds for investment in infrastructure projects in both developed and developing countries. It has been traditional for governments to own, operate and finance nearly all infrastructure services, primarily because their production characteristics and the public interest involved have been thought to require monopoly, and hence government provision. Funds raised from taxation have provided all or part of the public finance required for infrastructure projects such as roads, bridges, tunnels, telecommunication systems, water supply systems and power plants. As a result, a large proportion of national output in both developed and developing countries is spent on the provision of infrastructure services.

Developed countries have a large tax base provided by their large economies and facilitated by industrialisation. Infrastructure services have been relatively well developed through public finance strategies. Public finance has provided most of the major infrastructure projects procured since the end of the Second World War. In recent years, economic realities have led to the need to augment available public finance with private funds in an attempt to keep pace with infrastructure development as the population aspires to higher levels of infrastructure services.

In developed countries such as the UK, political motivation was the main driving force that enabled private finance to become an acceptable procurement method for major infrastructure projects over the last decade. However, the use of private finance instead of public finance for a particular project is only justified if it provides a more cost-effective solution. The financial plan of a project will often have a greater impact on its success than will the physical design or construction costs. The method of revenue generation, the type of project and its location often determine the cost of finance and its associated components. Government support to lenders of project finance, such as in the form of guarantees, has often been sufficient to ensure completion of projects that would not have been commercially viable without such support.

General government spending in most countries is financed primarily from direct and indirect taxation, concessions and royalties. There is increasing realisation that money is not free, even to governments who borrow from the private sector and bear the costs of collecting money themselves. The growing pressure on government budgets has led to a greater need and search for efficiencies in the financing and procurement of public infrastructure projects. This has led to the old method of financing and managing infrastructure by the private sector. These considerations have resulted in private finance being considered for a number of major projects normally procured with public funds, both in developed and in developing countries.

Developing countries generally lag behind in infrastructure development. Although there is a great need for new infrastructure projects, developing countries often lack funds from the normal sources expected in the developed countries. This is because developing countries are by definition poor. Public funds raised from taxation are inadequate to finance the required infrastructure projects. The inadequacy of public funds is mainly due to the low tax base in developing countries caused by the relatively weak domestic economies with low levels of industrial and commercial investment. Funding for projects is therefore scarce, loan finance difficult to obtain, and other necessary resources scarce. As a result, there is still a large gap between the demand and supply of infrastructure services. The need for private finance is even greater in developing countries where provision of infrastructure services is a major challenge and public finance from taxation is low.

Global political and economic considerations are also driving the move back to private financing of public infrastructure projects. With the gradual demise of exchange controls, domestic financial markets are being integrated into one global market. These markets have paved the way for funding projects all over the world. A project promoter may no longer need to concentrate on domestic lenders, bilateral lenders or multilateral lenders as sources of raising capital. Both domestic and international financial institutions are responding to promoters' needs. Tailor-made finance packages are being offered by financial institutions to finance projects. Multilateral and bilateral financial institutions that in the recent past provided the bulk of public sector funding for infrastructure development in developing countries now favour and demand the involvement of the private sector in the financing and management of infrastructure projects.

Successful infrastructure projects have also been financed through innovative public–private partnerships. In such cases, low interest rate loans or deferred or subordinated loans are provided by government in conjunction with loans from the private sector for major infrastructure projects, resulting in a hybrid mixed funded finance package.

One of the ways through which private financing of infrastructure projects is achieved, is by the use of project finance. One of the basic features of project financing is limited or non-recourse financing, in which investors in the project have no recourse to the general assets of the project promoter, but rely only on the project's assets and revenues. This limits the financing options because of the associated risk. However, the revolution in corporate financing techniques has opened new avenues. A host of financial instruments such as bonds, convertibles, warrants, depository receipts, export credits, lease finance and venture capital is now available to finance projects. Through financial engineering techniques, it is possible to tailor the financing needs of a project with instruments compatible with a project's projected cash flows.

Public financing is, however, seen as an easy method of procuring infrastructure projects, and has therefore been the traditional and conventional approach to infrastructure development in most countries. This book will concentrate on the more unconventional but fast developing method where private finance is utilised. Issues surrounding the use of private finance for infrastructure development are presented and discussed.

Tony Merna and Cyrus Njiru

Acknowledgements

The following experts contributed material that was used in the preparation of this book: Raghvendra Dubey, Gareth Owen, Steven F. L. Khu, Cyrus Njiru and Tony Merna. The authors wish to thank the external contributors for their expert contributions, without which production of this book would not have been possible.

The authors also wish to thank past and present members of the Centre for Research in the Management of Projects at UMIST for their assistance in the production of this book, including the following: Mr Rik Joosten, Mr Jozef De Groof, Mr Peter Hodgson, Mr Michael Methley, Mr John Carson, Dr Dimitrios Hatzis, Mr Richard Edwards, Ms Ruth Flynn, Dr Andras Timar, Mr Mark Elsey, Mr Philip Hurst, Mr Dougald Middleton, Mr Alasdair Sauders, Mr Paul Jobling, Mr Richard Shadbolt, Mrs Elizabeth Jenkins, Mr Timothy Hill, Dr Brian Winfield, Mr Michael D. Cannon, Mr Darrin Grimsey, Mr Peter Mullinger, Mr Mark Catchpoole, Dr Sandor Krupanics, Mr Emlyn Wit, Ms Helen Payne, Mr Douglas Lamb and Professor Nigel Smith.

List of contributors

Raghvendra Dubey, BA, MA, MSc, graduated from Patna University with a degree in economics and then worked with the Indian Ministry of Home Affairs. He later undertook a masters degree in economics at Rajasthan University and on graduating joined the Indian Ministry of Finance. He was Deputy Commissioner in the Department of Agriculture until September 1997. From 1997 to 1998 he studied for, and gained, an MSc degree in Management and Implementation of Development Projects from UMIST, where he concentrated his research work on financial engineering in the procurement of projects. He is now a director of the Ministry of Labour and based in New Delhi.

Gareth Owen, BEng, MSc, is a graduate of UMIST. After completing a degree in civil engineering he undertook an MSc in Engineering Project Management at the Centre for Research in the Management of Projects at UMIST. Gareth's research work concentrated on the aspects of project finance and its implementation in private finance initiative projects. Gareth spent several years working in the Structured Finance business at NatWest Bank, where he worked on a number of transport and infrastructure projects in the UK and overseas. Whilst there he was involved in arranging senior debt facilities for consortia via structured bond issues and syndicated loan facilities, as well as providing structuring advice to private sector clients. In addition he acted in an advisory capacity to public sector entities on structure, execution and negotiation. Gareth now works in the Structured Capital Markets Division of Barclays Capital.

Steven F. L. Khu, BSc, graduated from UMIST with a degree in Construction Management and is currently undertaking a doctorate programme at the Centre for Research in the Management of Projects at UMIST. His research is in the project finance field, specifically investigating the financial risks in projects and methods of restructuring the financial aspects of projects.

Cyrus Njiru, BSc Civil Eng (Hons), MSc, PG Dip San Eng (Delft), CEng, MICE, MCIWEM, FIEK, REng, is based at the Water, Engineering and Development Centre (WEDC), Institute of Development

Engineering at Loughborough University in the UK. He is currently involved in research, training and consultancy relating to the planning, provision and management of infrastructure services. He has considerable practical experience in infrastructure development and management, having held senior engineering and management positions in industry for over 16 years.

After graduating with an honours degree in Civil Engineering, Cyrus obtained initial field experience and training with consultants in Kenya and Japan. He subsequently studied Sanitary Engineering at the International Institute for Infrastructural, Hydraulic and Environmental Engineering (IHE), Delft, The Netherlands. He also pursued MSc courses in Engineering Project Management at UMIST, and in Management and Implementation of Development Projects at the Institute for Development Policy and Management (IDPM), Manchester University, both in the UK. This was followed by additional experience in design, construction, operation, maintenance and management of infrastructure projects while working closely with consultants, contractors, the public sector and international lending institutions. He also undertook postgraduate training in Project Finance and Public–Private Partnerships in Washington, DC, USA. He has undertaken field visits in several countries in Africa, Asia, Europe and the USA and studied many infrastructure projects. His research interests relate to aspects of the financing and management of infrastructure services in the context of developing countries, on which he is writing his PhD thesis at Loughborough University.

Tony Merna, BA, MPhil, PhD, CEng, MICE, MAPM, MIQA, is senior partner of Oriel Group Practice, a multidisciplinary research and consultancy organisation based in Manchester, UK. Tony is also a part-time lecturer at the Centre for Research in the Management of Projects at UMIST.

Contents

CHAPTER ONE

Project finance as a tool for financing infrastructure projects

Introduction

Development of infrastructure facilities in the industrialised countries was, until the beginning of the twentieth century, largely funded by private capital. This trend, however, changed and during most of the twentieth century, funds from the public exchequer have dominated infrastructure financing.

The provision of infrastructure facilities in the developing world, until the beginning of the twentieth century, was rudimentary. Most of these countries were under some form or other of colonial dominance. The colonial rulers developed limited infrastructure facilities primarily to meet their administrative requirements. No effort was made to develop social infrastructure facilities with its primary objective of meeting the needs of society. Soon after the end of the Second World War, when most of these countries gained independence, there was a sudden increase in the demand for infrastructure facilities. Due to the worldwide trend in public funding of infrastructure and the high cost of developing infrastructure facilities, the newly formed governments of these countries took on the responsibility of infrastructure development. The non-existence of capital markets in these countries also meant that private funding was not an option. The governments in the developing countries relied on fiscal measures (tax and non-tax) and external financing, through official development assistance, to fund many of these projects.

Since the early 1980s there has been growing realisation of the limitations of public funding for infrastructure development, in both the industrialised and the developing countries. Besides problems of accountability and efficiency often leading to high cost of provision for the consumers, public funding with its associated political considerations invariably led to poor performance and uneconomical pricing, which places severe strains on government budgets. The pressure on government budgets worldwide is increasingly leading to the adoption of private funding for infrastructure projects. One of the ways in which this is being achieved is through project financing.

Project financing is not a new tool. It has gained importance because the concept has gradually evolved as a specific financing technique in which the project lenders look only at the cash flows and earnings of the project as the source of funds for repayment of their investments, and not at the creditworthiness of the sponsoring entity. This has opened a number of avenues for the funding of new ventures that have no track record.

The logic for privately financing infrastructure projects is simple. In most cases, such projects would be severely delayed or perhaps would never be implemented if they were to wait for public financing from tax receipts.

In project financing, a project is considered as a distinct entity, separate from the promoter. The project does not, therefore, substantially impact on the promoter's balance sheet or the creditworthiness of the sponsors. Project financing is also known as non-recourse or limited recourse financing. The relationship among the various parties in project financing is established through a variety of contractual arrangements.

Rationale for investment in infrastructure

A large proportion of investment both in developed and developing countries is for the provision of new infrastructure. Developing countries invest about US $200 billion a year, that is about 4% of their national output, a fifth of their total investment and 40–60% of the public investment, on infrastructure. In high-income countries, value added to the infrastructure sector accounts for about 11% of their gross domestic product (GDP). The situation with regard to the availability of infrastructure facilities in developing countries, however, remains poor. Over 1 billion people living in developing countries do not have access to safe drinking water, 2 billion people lack adequate sanitation, electric power is yet to be accessible to about 2 billion people and demand for telecom facilities to modernise production facilities far outstrips the existing capacity. Rapid population growth is continuously increasing the demand for infrastructure. Although the precise linkages between infrastructure and development are still open to debate, there is no doubt that the existence of a decent level of infrastructure services helps determine one country's success and another's failure in diversifying production, expanding trade, coping with population growth, reducing poverty or improving environmental conditions. Good infrastructure raises productivity and lowers production costs, but it has to expand fast enough to accommodate growth.

The change from public financing to project finance

During the past 40 to 50 years, most of the infrastructure projects around the world, both in developed and in developing countries, have either been funded by the public exchequer or through a combination of public funds and foreign assistance. Funds raised by the public exchequer are either from fiscal measures (taxes, duties, tolls, fees, royalties) or loans raised from the private sources (internal as well as external). For the money raised through the fiscal measures, governments have to provide adequate justification to the affected population and also bear the cost of collection. Similarly, for the money raised through loans, adequate provision has to be made for servicing the loan. This puts a considerable burden on governments, both financially and administratively.

Besides these means, governments also resort to deficit financing through monetisation, which has a direct effect on inflation and macroeconomic balances. Governments have therefore been bearing virtually all the risks associated with infrastructure financing.

Large and varied impacts of infrastructure on development represent a very strong public interest and therefore the attention of government. In many countries, however, public predominance in infrastructure provision has not always been the case. It grew during the second half of the twentieth century although private participation in infrastructure was quite important in the nineteenth century and the first half of the twentieth century. The trend changed in the second half of the twentieth century when governments or parastatals became the overwhelming provider of infrastructure facilities, largely through vertically integrated, monolithic entities. The gradual increase in the dominance of the public sector in the provision of infrastructure services was primarily because of the recognition of the economic and political importance of infrastructure. There was a belief that problems with the supply technology required a highly activist response by governments, and a faith that governments could succeed where markets appeared to fail. This argument is also supplemented by the fact that in the case of most of the developing countries the government is the most creditworthy entity and is able to borrow at the lowest rates, making possible infrastructure projects that might not otherwise be financially viable.

Many countries made significant progress in the provision of infrastructure facilities under public leadership. It is, however, now being universally recognised that this situation has also resulted in serious and widespread misallocation of resources and poor performance and has failed to respond to demand. As stated earlier there are huge unmet demands of infrastructure facilities in most of the developing world. At the same time, the pricing policies for the publicly funded infrastructure services have strained the budgets of many governments. Many governments are facing acute macroeconomic problems and are therefore being forced to tighten their budgets, limiting further investment in development and maintenance of infrastructure facilities. For example, in the Philippines, public investment in infrastructure fell from 5% of GDP between 1979 and 1982 to less than 2% during the remainder of the 1980s. Such a sharp cut in investment in infrastructure may be the requirement for short-term macroeconomic adjustment, but is definitely not sustainable in the long run if higher economic growth rate is to be achieved. A high economic growth rate presumes an accompanying high investment in infrastructure.

A growing realisation of the above process is now leading to a shift in the role of government from provider of infrastructure facilities to facilitator of infrastructure development by encouraging private entrepreneurs and lenders to take a more direct role in the provision of these facilities. Private financing is needed not only to ease the burden on government finances but, more importantly, to encourage better risk sharing, accountability, monitoring and efficiency in management of infrastructure services. The challenge for the future is to route private savings directly to private risk bearers who make long-term investment in infrastructure projects. One method to achieve this end is the adoption of project finance techniques for the development of infrastructure facilities. This, in many developed and developing countries, requires far reaching reforms in legal and financial structures. These reforms will, however, bring in efficiency not only in infrastructure financing but also in the general development of capital markets.

Definition of project finance

The concept of project finance is widely used in business and finance in developed countries. Many developing countries are also using project finance to raise funds for their infrastructure projects. There is, however, no precise legal definition of project finance as yet.

The term 'project finance' is used to refer to a wide range of financing structures. However, these structures have one feature in common – the financing is not primarily dependent on the credit support of the sponsors or the value of the physical assets involved. In project financing, those providing the senior debt place a substantial degree of reliance on the performance of the project itself.

Nevitt (1983) describes the term 'project finance' as:

> Financing of a particular economic unit in which a lender is satisfied to look initially to the cash flows and earnings of that economic unit as the source of funds from which a loan will be repaid and to the assets of the economic unit as collateral for the loan.

Similarly, Merna and Owen (1998) have described project finance as:

> Financing of a stand-alone project in which the lender looks primarily to the revenue stream created by the project for repayment, at least once operations have commenced, and to the assets of the project as collateral for the loan. The lender has limited recourse to the project sponsors.

Thus, project finance is a useful financing technique for sponsors who wish to avoid having the project's debt reflected on their balance sheets or to avoid the conditions or restrictions on incurring debt contained in existing loan documents, or when the sponsors' creditworthiness or borrowing capacity is less than adequate. The cutting edge of developments in project finance is in emerging countries and to a large extent relates to private infrastructure projects as governments in these countries grapple with limited foreign exchange reserves, budgetary restraints and burgeoning infrastructure requirements.

Merna and Owen (1998) have described the concept of project finance with specific reference to a build–own–operate–transfer (BOOT) project as:

> Each project is supported by its own financial package, and secured solely on that project or facility. Projects are viewed as being their own discreet entities and legally separate from their founding sponsors. As each project exists in its own right, Special Project Vehicles (SPVs) are formulated. Banks lend to Special Project Vehicles on a non- or limited recourse basis, which means that loans are fully dependent on the revenue streams generated by the Special Project Vehicle, and that the assets of the Special Project Vehicle are used as collateral. Hence, although there may be a number of sponsors forming the Special Project Vehicle, the lenders have no claim to any of the assets other than the project itself.

Difference between project finance and corporate finance

It is important to understand the difference between project finance and corporate finance. Corporate finance is traditional finance where payment of loans to lenders comes from the organisation, backed by the organisation's entire balance sheet and not from a project alone. Lenders tend to look at the overall financial strength or balance sheet of an organisation as a prerequisite to lending for a project. So, even if a particular project fails, the lenders will still remain confident of being repaid because the organisation owning the project has a strong financial base. This does not mean, however, that the project will not be appraised for economic viability. A typical example of corporate finance is finance used to develop offshore drilling platforms owned by oil companies, who in most cases have healthy balance sheets.

In project finance, because the project is undertaken by a special project vehicle (SPV) and is an off-balance sheet transaction, lenders can expect significant losses if the project fails, because they do not have any recourse to the main organisation's assets. The only assets they have recourse to are the facilities of the project.

Basic features of project finance

From the above brief descriptions, the following basic features of project finance may be identified.

Special project vehicle

The first step in project financing is the setting up of an SPV, as a separate company from the promoter's organisation, which operates under a concession, normally granted by government. The sponsors of the project company usually provide the seed equity capital for the SPV. The SPV is usually highly geared (high debt/equity ratio).

Non-recourse or limited recourse funding

In non-recourse funding the lenders to the project, both debt as well as equity, have no recourse to the general funds or assets of the sponsor of the project. However, in limited recourse funding, access to the sponsor's general assets and funds is provided if the sponsors provide a guarantee of repayment, but only for certain risks. It has both advantages and disadvantages. Advantage, because the limitation of the project company to only project-related activities gives confidence to the lenders that the project would not be burdened with losses or liabilities from activities unrelated to the project. It also helps to perfect the security interests of the lenders in the project company with a right to replace the project management team in the event of poor performance of the project or even to foreclose and sell the project (step-in clauses) to recover their interests in the project to the maximum possible extent. Its disadvantage could be that the investors are left with a partially completed facility that has little or no residual value. Lenders therefore have to act very cautiously and completely satisfy themselves that the project facility will be able fully to meet its debt and equity liabilities, and on top of that earn a reasonable margin of profit for the sponsors to retain their interest.

Off-balance sheet transaction

The non-recourse nature of project finance provides a unique tool to project sponsors to fund the project outside their balance sheet. This structure enables funding of a variety of projects that might not otherwise have been funded, particularly when the sponsors either:

- are unwilling to expose their general assets to liabilities to be incurred in connection with the project (or are seeking to limit their exposure in this regard), or
- do not enjoy sufficient financial standing to borrow funds on the basis of their general assets.

Sound income stream of the project as the predominant basis for financing

The future income stream of the project is the most critical element in any project financing. The entire financing of the project is dependent on an assured income stream from the project, since lenders and investors only have recourse to the income streams generated by the project once it is completed, and assets of the project that may or may not have any residual value. The project sponsors, therefore, have to demonstrate evidence of future income through various means such as a power sales contract for a power plant, a concession agreement for a toll road project allowing the collection of tolls, or tenant leases for a commercial real estate project.

A variety of financial instruments

A project finance operation may involve a variety of financial instruments. These may be broadly classified as debt, equity and mezzanine finance. A brief description of financial instruments is provided below, and a detailed discussion of different financial instruments is presented in Chapter 3.

Debt

The most important element in project finance is the raising of the debt capital. The main attribute of debt capital is a specified return for the lender over a specified period of time. In project financing, the return on debt capital is linked to the income flows of the project entity and is protected only against the assets of the project. There are a variety of debt instruments such as pure loans and non-convertible debentures.

Equity

The process of project funding on a project financing basis normally begins with the setting up of a particular project legal entity floated by the project sponsors, described by Merna and Owen (1998) as an SPV. The project sponsors provide the seed equity capital for this project company. Later on, the project entity may also raise equity funds from the general public to part finance the construction and early operation phase. Equity is, however, risk capital and is subordinate to debt in terms of charge over the assets of the company. It shares in the profits of the project and any appreciation in the value of the enterprise, without limitation. The return on the equity, however, is the first to be affected in the case of financial difficulties being faced by the project entity.

Mezzanine finance

There are some kinds of financial instruments that are primarily in the form of debt but also share some qualities of equity capital. They are generally referred to as *quasi-equity*. This includes bonds, convertible debentures, preferred stocks and other instruments with attributes of both debt and equity.

Contractors, suppliers and purchasers

In many project finance operations, the contractors provide funding either in the form of equity contribution or by extending credits. Similarly, suppliers of the equipment participate in project financing through suppliers' credit that is often supported by the export credit guarantee organisations of the suppliers' country. Most of the projects financed under project finance have specified purchasers, for example a power project may sell its product to a power distribution agency via a transmission company. The power purchaser can participate in the project financing by extending advances against future purchases of power. They may also contribute to equity capital.

Sureties

These are the contingent funds allocated by the guarantors of the loans, suppliers' credits and advances to the project company, to protect the interests of the lenders and investors against any financial losses.

Insurance

These are the resources allocated by the insurance companies to compensate for the losses of the project in case of casualties, such as fire and other insurable events.

A variety of participants

A project finance operation involves a variety of participants. It has private participants who play the major role, government that provides necessary incentives, and domestic as well as international investors and lenders (multi-lateral, bilateral, commercial).

Private participants

Private sector companies and other private participants (domestic as well as foreign) often play the central role in a project financed through project finance techniques. The involvement of the private sector is in almost every aspect of the project:

- as the primary sponsor of the project – this is normally in conjunction with a host government sponsor in the case of developing countries; the involvement of the host government in the project builds the necessary confidence, among the lenders and investors, of the project having full government support
- as the major party responsible for the construction and operation of the project
- as the major financiers of the project through private financial institutions and commercial banks
- as guarantors or other sureties for certain types of transactions
- as insurers
- as purchasers of the output.

Government

The role played by the host government is generally crucial for the success of projects financed through project finance techniques. Most of the developing countries suffer from relatively underdeveloped legal and financial systems. The project finance, however, depends on very elaborate legal and contractual systems. Whereas performance of all the agencies involved in a publicly funded project is easily monitored through a bureaucratic system, performance in project finance is ensured through contracts and penalties. The first task of the host government, therefore, is to ensure that the requisite legislative and statutory reforms (enabling legislation) are put in place to make the project finance a viable alternative to public funding of projects. Besides this, the host government may also participate in the project in various ways, such as:

- participate as co-sponsor of the project along with private sponsor
- contribute to the equity capital
- contribute to the loan capital

- provide certain types of guarantees, such as against political risk
- provide certain resources required by the project which may still be within government control, such as coal supplies for a thermal power plant
- purchase the output produced by the project, such as electricity
- provide certain fiscal incentives, such as tax exemptions, tax holidays and subsidies.

Foreign governments

In many projects financed through project finance techniques the involvement of foreign governments is important. They participate in the project in various ways:

- by providing bilateral loans to the project
- by supporting the loans provided by private lenders and suppliers' credits through national export credit guarantee agencies
- by purchasing the product of the project company, such as electricity
- by providing raw materials for the operation of the project.

Sometimes a second level of foreign governments may also become involved in the project. For example, for a land-locked country the port facilities of a neighbouring country can provide the transit route for the project operation, such as import of raw materials and export of project produce. These countries may or may not be directly involved in the project but their participation in providing infrastructure often becomes crucial. Such a transit country may, at times, demand specific remuneration for providing specific transit facilities for the project's activities, like port facilities, road facilities, rail facilities and transmission lines or in some cases annuities.

Multi-lateral agencies

In many projects financed through project finance techniques, multi-lateral financial agencies, such as the World Bank, the International Finance Corporation, the Asian Development Bank and the African Development Bank, become involved. The involvement of these agencies is generally complementary to the private financing of the project and provides a catalytic role. The wide experience and involvement of expert teams of these agencies, in financing infrastructure projects in developing countries, provides confidence to other lenders and investors in the project. These agencies normally participate by providing loans and equity and through co-financing with other multi-lateral agencies.

A variety of risks

In conventional financing methods, the lenders look not only at the prospect of the project becoming successful but also at the general creditworthiness of the project sponsors. The risks associated with a particular project are not considered very critical because the lenders have access to the general assets of the project sponsors. In project financing the borrower is usually an SPV. The SPV will usually not have any past history since it is created just to implement and operate a specific project. Further, the lenders and investors have no or limited recourse to the general assets of the sponsor company. Diligence in understanding the risks associated with a project and careful attention to how they are allocated among project participants is the key to successful project finance development, investing and lending. It is much easier to put money into an ill-conceived project than to pull it out. Careful reflection about how a proposed project is intended to work in both good and bad times is just as appropriate for a power plant (power contract revenues) in the developing world as it is for an airport terminal serving a leading US city (rental income from airlines and concessions). The evaluation of the risks of diminution and interruption of the future revenue stream is the central question around which project finance revolves.

A detailed discussion on the identification, analysis and management of risks is presented in Chapter 4.

A variety of contractual arrangements

Although a project company is unusual in that it is established to undertake a single project, there is nothing unusual about the identity of the parties involved. All companies have owners, lenders, suppliers and customers, and all have dealings with the government. The difference in the case of project finance lies in the overriding importance of the contractual and financing arrangements that exist between these various parties. These are more than a series of independent bilateral arrangements. The contractual arrangements are primarily to address and allocate every major identified risk associated with the project, to the party that is best able to appraise and control that risk. Risk sharing through contractual arrangements is crucial for successful project financing. At the heart of project financing is a contract that allocates risks associated with the project and defines the claims and rewards. While often the cause of delay and heavy legal costs, efficient risk allocation has been central to making infrastructure projects financeable, and has been critical to maintaining incentives to perform. Risks are divided not

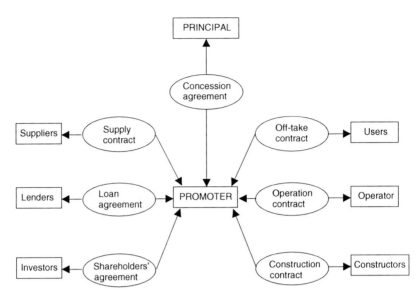

Figure 1.1. Project financing structure (Merna and Smith, 1996)

only between public and private parties but also among various private parties.

A simplified contractual arrangement structure for project financing is presented diagrammatically in Figure 1.1.

The importance of tax issues in project financing

Projects utilising project finance are often highly leveraged. They require large investments for long periods of time, sometimes up to 30 years. In most highly leveraged transactions, margins are very tight. The returns are acceptable only when measured against the low level of equity or the low level of risk. Because the margins are so low, the financing requires a high level of security and predictability of cash flow. Taxes, especially unplanned taxes, unacceptably reduce this cash flow. Thus, awareness of all potential tax implications throughout the life of the project is essential.

Although tax issues are country specific, there are many common themes, with each country's tax law being moulded by its level of development and sophistication. With very sophisticated projects, such as telecommunications, satellites and the like, the items of income and expense are not likely to be mentioned in the tax laws. It is of paramount importance that such issues are clarified through rulings or, in the case of a build–own–operate–transfer (BOOT) contract, in the concession agreement. Because multi-national teams are common, the various parties in a BOOT project are affected differently by the tax laws. Transnational tax issues often arise, adding layers of complication and the interplay of the tax laws of several jurisdictions.

Because tax laws are subject to change, it is common for the project documentation to include economic stability clauses. These clauses adjust returns, cash flow and/or revenues, to account for unanticipated changes in any of the host country taxes that may affect a project.

The major source of profit to a promoter organisation comes from operating a project. The tax on profits and other related taxes of contractors and suppliers, both in the host country and abroad, are relevant to the promoter as they affect the promoter's costs.

In many countries, title to assets of national importance cannot be vested in foreign corporations. They must generally be held by companies incorporated in the country where the project is located and such companies are often formed specifically for the project and are called *project companies*. Because of the size, risk and range of abilities needed, a consortium of companies representing the skills needed in different phases of the project is often formed.

The tax issues of the host country of the project company relate to taxation of:

- profits earned in the host country
- remittance of profits and taxation of other amounts paid to contractors, suppliers, investors and lenders
- gains, if any, when the project company is wound up or sold.

As previously stated, it is likely that the consortium will be a legal entity in the host country, although this does not necessarily mean it will be taxed as a legal entity there. It is also likely that the members of the consortium are incorporated elsewhere. Thus, there is the classic case of interplay between the host country and home country tax systems, and the possibility of several different home country systems.

The taxation of in-country operations by the host country can be reduced, if required, through the use of available tax holidays and accelerated depreciation methods. Developing countries often offer tax holidays during the early years of a project. As tax holidays may begin from either the first year or the first profitable year, it may be advantageous to maximise rather than minimise income from the project in the first few years. Delaying deductions and lengthening depreciation may be options to maximise the benefits of tax holidays.

Many jurisdictions limit depreciation to assets actually owned by the project company. If road connections, transmission connections or similar assets necessary for use in the project are built and given to the government (i.e. if they are not owned by the project company) depreciation is often not allowable. The costs must either be specially recovered under the project documentation or special tax treatment obtained. In some jurisdictions, the project assets become the property of the government immediately, even though they are subject to a concession for an extended period of time. Even if this does not affect the host country tax depreciation, it may have unplanned effects on home country taxation.

Related party payments for services, technology or interest on borrowings can be used to lower host country taxation. However, care should be taken, as they are most beneficial when the host country tax rate is not lower than the home country tax rate. Developing countries frequently limit the ability to make related party payments, so creative planning is often necessary.

Other potentially non-deductible amounts include investigation and bidding costs and foreign exchange losses. The first two can usually be resolved by attempting to take a home country tax deduction. Since the existence of exchange losses is often a financing rather than a tax issue, care should be taken to avoid them if they are non-deductible. Alternatively, they should be factored into the contract.

Once the host country tax has been determined, the appropriate structure for the investors should be determined. This encompasses a myriad of questions, which are complicated if the investors are from different countries. The fundamental problems include double taxation and the timing of tax payments. Issues include the existence of tax treaties between the host country and the investors' home countries and the home country system for taxation of foreign income. Among other things, tax treaties determine the level of withholding tax on remittances of profits, interest and royalties, and the application of capital gains taxes on the sale of the project company. Often treaties significantly reduce withholding taxes and waive capital gains taxes.

Should a treaty not exist between the host and home countries, then perhaps treaties exist through a third country. Ideally, this country will impose no or minimal taxes, forming a tax optimising treaty conduit. Developed nations frown on this approach and complicated legislation exists to discourage its use. Even in developing nations, convincing the tax authorities that one or more treaties apply may be difficult.

Equally as important is the home country tax system, particularly as it applies to the taxation of income earned overseas. How is double taxation – the application of host country and home country taxes to the same item of income – avoided? There are three general approaches to the question: worldwide taxation with credits, the similar tax system and the territorial system.

Some countries use a combination of the three. The *worldwide taxation system* brings all income earned anywhere into the home country's tax net. Double taxation is avoided through a foreign tax credit system. Home country taxes are reduced by foreign taxes paid. This may be viewed as a 'soak-up' approach, since the investor's tax liability is no less than if it had invested at home.

There is no long-term benefit to tax holidays and tax rates lower than the host country, since these benefits are soaked up by home country taxes. The *similar tax system* exempts foreign income that has already been subject to a foreign income tax from home country tax.

Finally, there is the *territorial system*, which exempts foreign income from home country taxation altogether. Investors from countries with the latter two systems benefit absolutely from tax holidays and lower tax rates. Investors from worldwide taxation countries will thus look very closely at intermediaries in tax havens to provide some benefit from the tax concessions on offer in projects.

Foreign tax credits are only allowed for foreign income taxes. Investors should ensure that the 'income' tax they are paying meets their home country definition of an income tax. Modifications of the income tax through the contract may render the income tax non-creditable.

Because some projects may incur start-up losses, the issue of loss flow-through can be important. As an example, the project company, although a legal entity in the host country, may be viewed as a partnership or other non-legal entity in the home country. This may allow home country tax relief for start-up losses and accelerated depreciation. Conversely, this type of structure results in immediate taxation of operating profits when they arise. To investors from worldwide taxation countries using the tax credit system, the benefit of start-up losses may offset the detriment of immediate taxation.

In several jurisdictions there are local income taxes in addition to national income taxes. In these situations the issue of allocation arises. Income earned in various local jurisdictions must rationally be allocated among the jurisdictions in order to avoid double taxation.

Tax treaties and foreign tax credits usually deal only with income taxes. Many of the taxes encountered are not income taxes. Turnover taxes under the heading of sales, use, value-added and business, amongst others, often dwarf the income taxes paid. Customs duties of 40% of equipment cost are not uncommon. With the capital-intensive nature of most major projects, customs duties can become one of the material costs in the project. Moreover, turnover taxes and duties are often paid from the beginning, long before any profits are made. Thus they are more in the nature of a project cost, whereas income taxes reduce the project's return since they only come out of profits.

An important but often overlooked tax issue involves the taxes associated with the transfer portion of a BOOT contract. Often these fall under the category of stamp duties or other transfer taxes. Care should be taken to avoid a taxable gain on the transfer. Although the last to be mentioned, these taxes are both the least understood and the most onerous. Included are VAT and sales taxes, the raft of taxes loosely called *turnover taxes*. As the name suggests, they are levied on the level of turnover, which generally has little to do with profits. VAT rates that are applied to gross revenue are often in the high teens. Getting such a figure wrong in a project could bankrupt the project company. Even more insidious are sales taxes, which lack the credit system of VAT (input VAT). Sales taxes compound, and apply at varying rates to items with little reason. There are also property taxes, stamp duties and many other schemes that need to be considered during the financial analysis of a project.

Tolls from the privately operated Dartford Crossing, Skye Bridge and Second Severn Crossing in the UK have recently been subjected to VAT imposed by the European Union. The promoters of these projects will not pay this additional tax, as it was identified as the responsibility of the UK Government, should it occur, under the terms of each of the concessions.

Infrastructure projects, particularly big projects in developing countries, often rely on imported equipment; both the equipment to build the project and the equipment which becomes part of the project. Duties and other imposts are usually levied on imports. The less developed the country, the higher the duties. Often equipment to be used on a construction project can be imported under bond, with the bond refunded when the equipment is re-exported. Developing countries often exempt from duties equipment that is to become a part of an infrastructure project.

Bibliography

Cannon, M. D. Taxation of BOOT projects. In Merna, T. and Smith, N. J. (eds). *Projects Procured by Privately Financed Concession Contracts*, vol. 2, 1st edition. Asia Law and Practice, Hong Kong, 1996.

Merna, T. and Owen, G. *Understanding the Private Finance Initiative*. Asia Law and Practice, Hong Kong, 1998.

Merna, T. and Dubey, R. *Financial Engineering in the Procurement of Projects.* Asia Law and Practice, Hong Kong, 1998.

Merna, T. and Smith, N. J. *Guide to the Preparation and Evaluation of Build Own Operate Transfer (BOOT) Project Tenders,* 2nd edition. Asia Law and Practice, Hong Kong, 1996.

Nevitt, P. K. *Project Finance,* 4th edition. Bank of America Financial Services Division, 1983.

Merna, T. and Smith, N. J. (eds). *Projects Procured by Privately Financed Concession Contracts,* vol. 2, 1st edition. Asia Law and Practice, Hong Kong, 1996.

CHAPTER TWO

Public finance for infrastructure projects

Review of methods used in financing infrastructure projects

Methods of financing infrastructure projects may be categorised as follows:

- public finance obtained from general taxation and public borrowing
- private finance
- public–private partnership (PPP) in financing.

Public finance is the traditional source of funds for investment in infrastructure projects in both developed and developing countries. Governments traditionally own, operate and finance nearly all infrastructure, primarily because its production characteristics and the public interest involved were thought to require monopoly, and hence government provision. Funds raised from taxation have provided all or part of the public finance required for infrastructure projects. Public finance is also used to subsidise existing infrastructure projects such as rail and air transportation.

Typical allocation of government funds

Developed countries have a large tax base provided by their large economies and facilitated by industrialisation. Infrastructure services have been relatively well developed through public finance strategies. In recent years, economic realities have led to the need to augment available public finance with private funds in an attempt to keep pace with infrastructure development as the population aspires to higher levels of infrastructure services. Government funds in the UK are allocated to the following different departments:

- social security
- miscellaneous (see below)
- health

- education
- defence
- debt interest
- law and order
- transport (see below)
- social services
- reserve and adjustments.

The miscellaneous category consists of housing and urban regeneration, local services, agriculture, finance and management, industry and science, employment, environmental management, international relations, heritage and leisure, and net EC contributions.

The transport category consists of local roads, trunk roads, national railways, local transport, London transport, aviation, the Drivers and Vehicles Licensing Association (DVLA) and sea transport.

In the UK, nearly a third of government spending is on social security with debt interest fluctuating, as this is dependent on the amount of money borrowed by the government to meet spending requirements. In 2000, £85 billion of government spending was raised from direct taxation. The top 10% of earners paid 50% of this amount. Of this direct taxation, 70% was paid by 30% of the taxpayers residing in the south-east of the UK. The UK has reduced debt interest over the last 4 years. In Germany, however, debt interest is currently DM 150,000 per minute on a debt of DM 400 billion, and still rising.

At the time of writing this book, the UK government spending for the next fiscal year (2002–2003) is forecast to be in the region of £400 billion. It is interesting to note that at the same time the London Foreign Exchange Market and Insurance Market are £638 billion and £1.8 trillion markets, respectively. The potential for private sector funding is evident.

Transportation projects in the UK have received only 3% of government spending over the last few years, the largest amount being spent on local roads. The funding by the UK government for transportation projects (£10.2 billion in 1996–1997) was supplemented by private finance through private finance initiative (PFI) or PPP programmes. This example shows that government budgets even in developed countries with a large tax base are strained and need to be augmented with funds from the private sector in order to finance needed infrastructure projects.

Developing countries generally lag behind in infrastructure development. Although there is a great need for new infrastructure projects, developing countries often lack funds from the normal sources expected in the developed countries. This is because developing countries are by definition poor. Public funds raised from taxation are inadequate to finance the required infrastructure projects. The inadequacy of public funds is mainly due to the low tax base in developing countries caused by the relatively small economies with low levels of industrial and commercial investment. Funding for projects is therefore scarce, loan finance difficult to obtain and other necessary resources scarce. Methods for funding projects in developing countries are discussed elsewhere in this book.

Sources of government funds

Most governments utilise an array of instruments for extracting money from the economy to finance public spending. These have been progressively introduced to provide sufficient income to meet spending plans and borrowing targets. Major instruments utilised in many developed countries such as the UK include:

- income tax
- National Insurance
- corporation tax
- value added tax
- fuel duty
- vehicle excise duty
- alcohol duty
- tobacco duty
- council/property tax
- business rates
- privatisation proceeds
- National Lottery tax
- 3 G telephone licences.

In the case of fuel duty in the UK, for example, the duty paid on each litre of fuel is approximately 80% of the retail price. Recent projections for the year 2002 suggest that fuel duty and vehicle excise duty will provide £50 billion to the government. However, only 20% of this sum will be used by government to finance road infrastructure projects. Until recently the UK government had a policy of raising the rate by at least 5% above inflation each year, for environmental reasons.

Other instruments include:

- customs duties
- stamp duty

- betting and gaming duties
- inheritance tax
- capital gains tax
- petroleum revenue tax
- insurance premium tax
- oil royalties
- air passenger duty.

A special type of tax, known as the *hypothecated tax*, has been introduced on a number of projects in Europe by governments. This tax is raised specifically for one project or one end use. A hypothecated tax was used, for example, in the Sparta Airport project in Athens, Greece, in the form of a tax on airline tickets to raise the required equity stake. Although this project was procured by private finance the government's equity stake was raised using this method of taxation.

Governments also borrow money to meet the public sector borrowing requirement (PSBR). This is mainly carried out by the issue of government stock, in the UK usually known as *gilts* (gilt-edge securities), typically with redemption dates of 10–20 years. Interest may be paid at a fixed rate or linked to an index, such as the retail price index (RPI). The other significant method of borrowing is National Savings. It should be noted that many projects in EU member states often received an element of their funding through EU grants or loans.

In the USA government bonds are called *treasuries*, because the Treasury Department sells them. The maturity dates of these bonds range from 3 months to 30 years. Treasury bonds are considered the safest investment in the world because they have 'full-faith-and-credit' backing of the US government. The monies raised from these bonds are often used to finance infrastructure projects.

Many countries also utilise instruments for raising finance specifically through tax levies on imported goods, for example cars and electrical goods.

The traditional argument, over and above political reasons, for government financing projects has been that governments can borrow money at a lower interest rate than the private sector. It should be noted that there is no such thing as free money. Governments need to employ, equip and accommodate large departments to administer the collection of duties and the management of loans. In many cases, low interest rate loans or deferred or subordinated loans were provided by governments in conjunction with loans from the private sector for major infrastructure projects, resulting in a hybrid mix funding finance package.

Much of the world's existing infrastructure, in market, planned and command economies has been financed by the public sector. Infrastructure has often been classified into social infrastructure and economic infrastructure. Economic infrastructure comprises the long-lived engineered structures, equipment and facilities, and the services they provide that are used in economic production and by households. This infrastructure includes utilities such a power, gas, electricity, telecommunications, water supply and waste collection and disposal, dams, canals, roads and other transport sectors, ports, waterways and airports. Social infrastructure primarily concerns services such as education and health care.

Although much of the infrastructure identified above has been financed by the public sector, many countries now seek to procure infrastructure through the private sector.

Methods of procuring projects

Where governments have been responsible for financing projects directly or through their agencies, they have, like the private sector, utilised an array of contract strategies or arrangements in an effort to meet budgeted costs and provide an equitable risk share. Strategies and arrangements include:

- fixed priced lump sum turnkey contracts
- management contracts
- own period tendering (OPT)
- lane rental
- design and build
- partnering.

In most cases competitive tendering procedures are used, the three main stages being pre-qualification, tender documents and evaluation of returned bids. Evaluation of bids must consider in great detail the tendered price, since in public sector funded projects a budget will normally have been set and in private sector projects, where finance is often raised from corporate bonds, sufficient monies are available to meet the cost of the project.

Governments have also looked at involving the private sector in the operation and maintenance of publicly funded projects, usually on the basis of a lease or concession for specific period of time.

Public financing is seen as an easy method of procuring infrastructure projects, and is therefore the traditional and conventional approach to infrastructure development in most countries. This book will therefore concentrate on the new, but fast developing, methods where private finance is utilised.

Bibliography

Cocks, R. and Bentley, R. *Government Spending: The Facts.* Databooks, 1996.

Merna, T. and Njiru, C. *Financing and Managing Infrastructure Projects.* Asia Law and Practice, Hong Kong, 1998.

Merna, T. and Smith, N. J. *Mechanisms for the Award of Competitive Tenders for Works and Supply Contracts.* Asia Law and Practice, Hong Kong, 1996.

CHAPTER THREE

Financial instruments

Introduction

All projects require financing and no project can progress without financial resources. However, the nature and amount of financing required during different phases of the project varies widely. In most engineering projects the rate of expenditure changes dramatically as the project moves from the early stages of studies and evaluation, which consume mainly human expertise and analytical skills, to the design, manufacture and construction of a physical facility. Broadly, a project may be said to pass through three major phases:

- project appraisal
- project implementation/construction
- project operation.

A typical cumulative cash flow curve for a project is illustrated in Figure 3.1.

The precise shape of the cash flow curve for a particular project depends on various factors, such as the time taken in setting up the project objectives, obtaining statutory approvals, design finalisation, finalisation of the contracts, finalisation of the financing arrangement, the rate and amount of construction and operation speed. Negative cash flow, until the project breaks even, clearly indicates that a typical project needs financing from outside the project until it breaks even. The shape of the curve also reveals that in the initial phase of the project relatively less financing is required. As the project moves on to the implementation phase there is a sudden increase in the financial requirements, which peaks at the completion stage. The rate of spending is depicted by the gradient of the curve. The steeper the curve the greater is the need for finance to be available.

Once the project has been commissioned and starts to yield revenues the requirement for financing from outside the project becomes less and less. Finally, the project starts to generate sufficient revenues for operation and maintenance and also a surplus. However, even after the break-even point, the project may require financing for short periods, to meet the mismatch between receipts and payments.

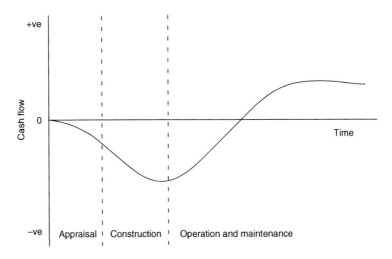

Figure 3.1. Typical project cumulative cash flow diagram

In project financing it is this future cash flow that becomes the basis for raising resources for investing in the project. It is the job of the finance manager of the project to package this cash flow in such a way (robust) that it meets the needs of the project and at the same time is attractive to the potential agencies and individuals willing to provide resources to the project for investment. In order to achieve this objective effectively a thorough knowledge of the financial instruments and the financial markets in which they trade is essential. In this chapter various financial instruments available for project financing are discussed.

Financial instruments

Projects have to raise cash to finance their investment activities. This is usually done through issuing or selling securities. These securities, known as 'financial instruments', are in the form of a claim on the future cash flow of the project. At the same time, these instruments have a contingent claim on the assets of the project, which act as a security in the event of future cash flows not materialising as expected. The nature and seniority of the claim on the cash flow and assets of the project vary with the financial instrument used. Merna and Owen (1998), describe financial instruments as the tools used by the special project vehicle (SPV) to raise money to finance the project. Traditionally, financial instruments were either in the form of debt or equity. Developments in the financial markets and financial innovations have led to the development of various other kinds of financial instruments, which share the characteristics of both debt and equity. These instruments are normally described as 'mezzanine finance'.

Debt instruments refer to those securities issued by the project that makes it liable to pay a specified amount at a particular time. Debt is senior to all other claims on the project cash flow and assets.

Ordinary equity refers to the ownership interest of common stockholders in the project. On the balance sheet, equity equals total assets less all liabilities. It has the lowest rank and therefore the last claim on the assets and cash flow of the project.

Mezzanine finance occupies an intermediate position between the senior debt and the common equity. Mezzanine finance typically takes the form of subordinated debt, junior subordinated debt, bonds and preferred stock or some combination of each.

Besides debt, equity and mezzanine finance a project may also utilise certain other types of instruments such as leasing, venture capital and aid.

Since the financing requirement of a project depends on the future cash flow, which in turn depends on time, another way of classifying financial instruments is temporal in nature; that is, long-term financing instruments and short-term financing instruments.

Short-term financing refers to financial instruments, which normally have a repayment obligation of up to 1 year. *Long-term financing* refers to raising of debt, equity and mezzanine finance, which have a repayment obligation beyond 1 year.

From a project financing point of view it is more reasonable to discuss the financial instruments according to their maturity (i.e. short term or long term).

Long-term financing instruments

A project raises long-term financing primarily for long-term investment purposes. Long-term financing is needed because the asset created by the project has a gestation lag before it starts to yield revenues. Long-term financing helps the project by deferring, partly or fully, the servicing of the securities sold until the project starts to generate revenues (Merna and Dubey, 1998).

Debt instruments

Debt instruments refer to the raising of term loans from banks, other financial institutions (including commercial banks, merchant banks, investment banks, development agencies, pension funds and insurance companies), debentures and export credits (buyers' and suppliers' credit). In terms of seniority, senior debt ranks the highest among the financial instruments in terms of claims on the assets of a project. This means that in the event of default the lenders of senior debt have the first right to claim the assets of a project.

Term loans

Term loans are negotiated between the borrower (project) and the financial institutions. In large infrastructure projects where the funds required to be raised are very high, term loans are usually provided by more than one bank or financial institution. A group of banks and financial institutions typically pool their resources to provide the loans to the project. For example, in the Channel Tunnel Project the original financing structure of this project involved 210 lending organisations, with 30% of the loan arranged through Japanese banks. The lead bank was NatWest Bank. Banks and financial institutions set their own internal exposure limits to particular types of project. This helps in spreading the risk. Generally, an investment bank or a merchant bank acts as the agent or lead bank to manage the debt issue. Many banks specialise in lending to certain types of infrastructure projects for which they possess both technical and financial experience.

The terms and conditions of loans vary from lender to lender and borrower to borrower. Loans can have a fixed interest rate or a floating interest rate. Repayment of the loan could be between 7 and 10 years for an oil sector project to between 16 and 18 years for a power project (Merna and Owen, 1998). It varies with the gestation lag of the project. The servicing of the loan, that is the repayment of the principal amount (amortisation) and the interest payment, can take various forms. It could be anything from equal instalments spread throughout the life of the loan to a bullet or sunset payment at the end of the period. Another alternative could be an initial moratorium on repayment of the principal amount but with regular interest payment. A third option could be a moratorium on both the principal repayment and the interest payment until the project is planned to operate. The project characteristics and the availability of the financial instruments determine the type of loan. As stated earlier, the success of the project lies in projecting attractive future cash flows, which are acceptable to lenders, since in project finance the basic security to the lenders is the cash flow of the project.

The cost of raising debt capital includes certain fees besides the interest. These are the:

- *management fee* – a percentage of the loan facility for managing the debt issue, normally to be paid up front

- *commitment fee* – calculated on the undrawn portion of the loan to be paid when the loan is fully drawn
- *agency fee* – normally an annual fee to be paid to the lead bank for acting as the agent to the issues after the loan has been raised
- *underwriting fee* – paid up front as a percentage of the loan facility to the bank or financial institution which guarantees to contribute to the loan issue if it is not fully subscribed
- *success fee* – paid up front as a percentage of the total loan once all loans have been secured
- *guarantee fee* – paid annually on the outstanding loan amount if it is guaranteed against default.

All, none or some of these may be present in a specific loan proposal. In certain cases the lenders submerge all the fees in the interest rate they offer. A careful analysis of the cost of loans offered from different sources is therefore required. The overall cost of raising a term loan is, however, still less than that for any other mode of large-scale financing because the project has to negotiate and deal with only a small number of lenders of money, through a lead manager of the issue. Also, in the event of default it is easier to renegotiate a term loan than it is any other instrument of financing.

Mobilising commercial debt

According to the International Finance Corporation (IFC), most projects are financed utilising debt as part of the financing package. In project finance, mobilising commercial debt can be quite difficult for several reasons:

- *High demand, cautious lenders.* Lenders face the same risks as equity investors when arranging loans to finance a project. They also fear the risk of not getting any money back in the event of default. In addition, there is a limit to how far loan pricing can be pushed. So, the lenders often have the major say in how financing is to be structured and seek to reduce the project's risks by negotiating the conditions with the promoter under which they will participate.
- *Foreign lenders.* In developing countries, most domestic markets cannot mobilise high volumes of long-term debt, so they turn to foreign lenders. Foreign lending involves foreign currency, thus exposing the promoter to currency risks.
- *A limited pool of lenders.* The number of banks involved in project finance is still relatively small, although more have entered the market in the last 5 years. Each bank has exposure limits to the

project finance environment, and thus organising a syndication of lenders is complex and time consuming.

- *Restriction on non-bank local lenders.* In developing countries there are many potential non-bank lenders such as pension funds and insurance companies. However, this potential is not being used, as many of these companies are publicly owned monopolies and most developing countries also require pension funds and insurers to invest mostly in government securities.

Subordinated debt

In project financing there is another type of debt called subordinated debt. Subordinated debt is that debt which is subordinate to senior debt, and generally only has second claim to the collateral of the project company. This means that in the event of default by the promoter, lenders of all senior debt must be paid before lenders of subordinated debt. As is it is second only to senior debt in terms of claims on the project's assets, lenders seek higher returns on subordinated debt. The interest rate on subordinated debt is usually higher than the interest rate on senior debt. For example, the interest rate on senior debt may be LIBOR + 200 BP, but the interest rate on subordinated debt can be LIBOR + 400 BP, where LIBOR is the London Interbank Offer Rate and BP is basis points. Subordinated debt is used mainly for refinancing needs or for restructuring of the finance package of a project.

Stand-by loans

Stand-by loans are often arranged with lenders by promoters to meet required draw downs in excess of the term loans, which are often due to a higher interest rate, lower than expected revenues in the early phase of the operation, and additional works in the construction phase.

Euro currency loans

Euro currency loans are the loans provided by banks from the unregulated and informal market for bank deposits and bank loans denominated in a currency other than that of the country where the bank initiating the transaction is located. For example, a British bank in London that lends in French Francs to a German company is in fact lending in Euro currency. Euro currency finance originated in the 1950s. The development of the Euro currency loan market coincided with the development of the inter-bank market for Euro currency deposits and placings. A Euro currency deposit can be defined as a deposit in a currency other than the currency of the country in which the bank is located.

The Euro currency market provides a lot of flexibility in borrowing by offering funds with fixed or floating interest rates. These loans are generally for a fixed period. When fixed-rate funds are scarce or too expensive for the borrower, the lending bank provides commitment to lend for, say, 5 years on a rolling basis, where the rates are adjusted at agreed intervals (monthly, quarterly, semi-annually) in line with the then current market rate. Another feature of Euro currency finance is the multi-currency credit facility. This enables the borrower to draw down in any currency of his choice, although this condition is subject to the availability of funds in the currency of choice.

Banks usually provide these funds out of their Euro currency deposits, which banks take for a short term for major currencies up to 1 year and in the case of US dollars, Sterling and Deutschmarks up to 5 years, depending upon the amount. (Note: from January 2002 the Deutschmark will be replaced by the Euro.) The loans are provided on a matching basis with deposits procured. The largest sector of the Euro currency market is deposits and loans in US dollars and therefore it is also known as the *Eurodollar market*. London is the principal source of Euro currency finance.

Debentures

A debenture is similar to a term loan except for the fact that the loan is divided into securities and sold through the stock market to a variety of investors. Debentures are usually in the form of a bond, undertaking the repayment of the loan on a specified date, and with regular stated payments of interest between the date of issue and date of maturity. In the UK, the Companies Act defines the word 'debenture' as including debenture stock and bonds. The terms 'debenture' and 'bond' are often used interchangeably in the UK. However, in this book debentures and bonds are treated as two different financial instruments for two main reasons. Firstly, bonds are normally treated as subordinate to debentures in the event of liquidation of the borrowing entity. Secondly, because bond markets have become much more flexible through the introduction of a variety of bond instruments, bonds have become very close to equity issues as far as liquidity is concerned. *Liquidity* refers to the ease and quickness of converting assets to cash. It is also called *marketability*. Bonds have therefore been classified as part of mezzanine finance instruments.

Debentures can be issued to the public only when a detailed prospectus accompanies them. Debenture stockholders normally have a trustee acting on their behalf to check that the project is adhering to the provision laid down in the trust deed. For example, the project pays the interest and principal as and when stated. The trustee is normally an insurance company or professional firm, which charges a fee for acting as a trustee.

In the UK, a debenture is usually secured against the company's assets, either with a fixed charge, a floating charge or both. A fixed charge or mortgage debenture is fully secured against the assets of the project; that is, a particular real asset of the project has a charge placed upon it. Creditors have full security in this case. This, however, puts a restriction on any proposal to dispose of that asset until there is a charge on it. All the debenture holders are required to be consulted, which is often difficult.

Debenture with a floating charge has charge attached to all the present and future assets of the project without any particular asset being specified. The project is free to deal with the assets without referring to the debenture holders. However, if the project defaults or is wound up, the floating charge crystallises and becomes a normal fixed charge on all the present and future assets of the project.

Sometimes unsecured debentures are issued, which have no security. The debenture holders have a note of indebtedness from the project. In the event of liquidation of the project the unsecured debentures are not paid until the secured debentures have been paid.

In the USA, a debenture always refers to an unsecured bond. At the current time, almost all bonds issued to the public by industrial and finance companies in the USA are debentures. However, most utility and railroad bonds are secured by a pledge of assets. Further, whereas in the UK in a debenture 'stock' refers to debt security, the term refers to equity shares in the USA.

Except for unsecured debentures, which are rare, other forms of debentures require security of the assets of the project. Since the method of project financing primarily addresses the needs of projects, which have no or few assets to begin with, raising funds through the issue of debenture is uncommon. However, after assets have been created this method can be used to raise resources. Except in the form of bonds debentures are not common instruments in financing projects.

Export credit

Export credit is a pure form of loan provided by either the exporter of the product or equipment or by a bank in the exporting country or by any governmental agency in the exporting country set up to promote exports. These loans are guaranteed by the export credit guarantee agencies in the exporting countries, such as the:

- Export Credit Guarantee Department (ECGD) in the UK
- Export Development Corporation (EDC) in Canada
- Compagnie Française d'Assurance pour le Commerce Exterieur (COFACE) in France
- Kreditanstalt fur Wiederaufbau (KfV) in Germany
- Sezione Speciale per l'Assicurazione del Cedito all'Esportazino (SACE) in Italy
- Ministry of International Trade and Industry (MITI) in Japan
- Private-Export Funding Corporation (PEFCO) and Overseas Private Investment Corporation (OPIC) in the USA.

The guarantee fee depends largely on the political risk perception about the borrowing country. For example, during the Gulf Crisis of 1991, when India faced acute balance of payment problems and there was heavy risk of default, most of these agencies raised their insurance premium to 16–18% up front as against the normal premium of 4–5% on a 5-year loan. Export credit is either in the form of supplier's credit or buyer's credit.

Supplier's credit is normally structured in one of two ways:

- If the supplier of the machinery or equipment is in a position to sell on a deferred credit basis, it enters into an agreement with the importer to supply the equipment and receive payments for it over a period of time. There is no involvement of any bank or financial institution in this structure. The supplier gets the deferred credit insured from the export credit guarantee agency of the country.
- If the supplier is unable to extend deferred credit, it negotiates with a bank or financial institution or governmental export promotion agency to seek a loan for the export with a guarantee from the export credit guarantee agency. The exporting company takes the loan in its own name and extends deferred credit to the importer as above. This is called suppliers' credit because the supplier of the equipment extends credit to the importing project. The importer of the equipment seeks a loan from a bank or financial institution in the exporting country, guaranteed by the export credit guarantee agency of the exporting country, and makes the payment to the supplier of the equipment. In this structure it is the liability of the importer to repay the loan to the lending agency.

Ordinary equity

As stated earlier, the process of a project proposed to be funded on a project financing basis starts with the setting up of a particular project legal entity, known as an SPV. The sponsors of the project provide the initial equity capital, known as the *seed equity capital.* Merna and Owen (1998) define equity capital, or *pure equity,* as the provision of risk capital by investors to an investment opportunity. This usually results in the issuance of shares to those investors. A share may be described as an intangible bundle of rights in a company, which both indicates proprietorship and defines the contract between the shareholders. The terms of the contract, that is the particular rights attaching to a class of shares, are contained in the article of association of the company. Equity is the residual value of a company's assets after all outside liabilities (other than to shareholders) have been allowed for. Equity is also known as *risk capital,* because these funds are usually not secured and have no registered claim on any assets of the business, thus freeing these assets to be used as collateral for the loans (debt financing).

Equity, however, shares in the profits of the project and any appreciation in the value of the enterprise, without limitation. The compensation for equity is dividends (dividends are the amount of profits paid to shareholders). No dividends are paid if the business does not make profits. Dividends to the shareholders can be paid only after debt claims have been met. The return on the equity, therefore, is the first to be affected whenever financial difficulties are faced by the project entity. This means that equity investors, in the worst-case scenario, may be left with nothing if the project fails, and hence they demand greater return on their capital in order to bear a greater risk. This explains the general rule where high-risk projects use more equity while low-risk projects use higher debt.

A high proportion of equity means a low financial leverage and a high proportion of debt equals a high leverage. *Leverage* is measured as the ratio of long-term debt to long-term debt plus equity. Leverage is also called *gearing* or the *debt/equity ratio.*

High financial leverage means that relatively more debt capital has been used in the project, signifying more debt service and less funds being available for distribution to the equity holders as dividend payments. However, once the project breaks even and profits start to grow rapidly, shareholders receive a higher dividend. The seed capital provided by the sponsors of the project, which is normally a very small amount as compared to the total finances raised for the project, is also known as *founders' shares* or *deferred shares.* These are lower in status than ordinary and preference shares in the event of winding up.

In non-recourse financing the debt/equity ratio may be higher if the interest rate is high, provided lenders are satisfied with the risk structure of the project. If, however, a project is considered innovative then lenders will demand more equity and the equity will be drawn down before debt becomes available to the project.

Ordinary share capital is raised from the general public. Holding of these shares entitles dividends, which provide the right of one vote per share held, and the right to a *pro rata* proportion of the project's assets in the event of winding up of the project. The right to participate in the assets of the project provides the opportunity for the highest return on the capital invested.

The SPV can raise equity capital from the market in various ways, such as by public issue, offers for sale, issue by tender, private placement and rights issue.

Cost of equity issue

The detailed procedures (public offer, offer for sale, issue by tender, private placement) involved in getting a new issue registered, issuing a detailed prospectus and aggressive marketing through advertisements make it very expensive. On top of that the issuer has to pay underwriting commission, a Stock Exchange listing fee, legal and broker's fees and capital duties. It is estimated that for a public offer the total cost is as high as 12.5% of the proceeds for a £5 million share issue.

Mezzanine finance instruments

There is a host of financial instruments in this category. Mezzanine instruments are senior with respect to an equity issue and subordinate to debt. Some of them are similar to a debt issue and some of them share features of an equity issue.

A bond, like any other form of indebtedness, is a fixed income security. The holder receives a specified annual interest income and a specified amount at maturity – no more, no less (unless the company goes bankrupt). The difference between a bond and other forms of indebtedness such as term loans and secured debentures is that bonds are subordinate forms of debt as compared to term loans and secured debentures. Similar to debentures these are issued by the borrowing entity in small increments, usually US $1000 per bond in the USA. After issue, investors on organised security exchanges can trade the bond. Four variables characterise a bond: its par value, its coupon, its maturity and its market value.

Par value

Par value is also known as *nominal value, face value, principal* or *denomination*. It is stated on the bond certificate. It is the amount of money which the holder of the bond will receive on maturity.

Coupon rate

The coupon rate is a percentage of the par value the issuer promises to pay the investor annually as interest income. This is normally paid semi-annually. Bonds are normally issued with attached coupons for the payment of the earned interest. The holder of the bond in such cases separates the coupons and sends them to the company for the payment of interest. This is the reason why interest in the case of bonds is known as the coupon rate.

Maturity date

The date on which the last payment on the bond is due is known as the maturity date.

Market value of bond

On the date of issue of the bonds the issuer usually tries to set the coupon rate equal to the prevailing interest rate on other bonds of similar maturity and quality. This ensures that the initial market price of the bond is almost equal to its par value. After issue, the market value of the bond can differ substantially from its par value. Since the stated return on the bond (its coupon) is fixed, the market value of the bond depends on the general level of interest rate prevailing in the market. As compared to the coupon rate, if the general interest rate in the market is low then the price of bond moves up, since holding of the bond yields a higher return as compared with other forms of investment in the market. On the other hand, if the general rate of interest in the market is high as compared to the coupon rate of the bond, the price of the bond will be quoted low, since there will be less demand for these securities. In the example illustrated below the bond has been issued with a par value (redemption value) of one, with a coupon rate of 6.5% and maturing on 22 May 1996. If the bond is purchased today at the current market price of 0.825 it will mean the effective coupon is 7.9%, and if the instrument is held until maturity (yield to maturity (YTM)) then it will yield a revenue of 11.1%.

Market price:	Redemption:	Maturity:	Coupon:	Current yield:	YTM:
0.825	1.00	22 May 1996	6.5%	7.9%	11.1%

Issuing of bonds

A project entity can raise finance through the issue of bonds in two ways: public issue and private placement. Public issue is described below.

Public issue

The procedures required to be followed for public issue of bonds are broadly the same as those for a public equity issue. First of all the public offering should be approved by the board of directors of the SPV. Sometimes a vote of the equity holders is also required. Thereafter, the SPV should file a registration statement with the designated authority, such as the Securities & Exchange Commission (SEC) in the USA, the UK Registrar of Companies in the UK and the Security and Exchange Board of India (SEBI) in India. This statement contains a great deal of financial information, including a financial history, details of existing business, proposed financing and plans for the future.

After the issue has been registered the securities can be sold. A selling effort through the issue of a detailed prospectus is made. The prospectus is advertised in national dailies for the information of the general public. The advertisement contains an invitation to the public to apply for the bonds and a closing date for applications is specified.

Unlike an equity issue the registration statement of a bond issue includes an *indenture*, which is a written agreement between the borrower and a trust company. It is also known as the *deed of trust*. The trust company is appointed by the borrower to represent the bondholders. It is the responsibility of the trust company to ensure that the terms and conditions of the indenture are adhered to by the borrowing entity. A typical bond indenture includes the following provisions:

* the basic terms of the bond
* security or guarantees
* covenants
* a sinking fund arrangement
* a call provision.

Each of these provisions is discussed below.

The basic terms of the bond

These are the par value, coupon rate and maturity date. These terms have been explained earlier in this chapter. The trust company makes sure that the coupons and repayments of the principal are paid regularly by the borrower. In the event of default on payment by the borrower the trust company represents the holders of the bond.

Security or guarantees

Bond issues have normally some asset of the borrower pledged as security for repayment and coupon payment. This asset is known as *collateral*. In the case of project financing, since there is very little or no commensurate asset to act as collateral, the borrower obtains a guarantee for the servicing of the bond issue. An example is the Road Management Group project in the UK. The project was to design, build, finance, operate and maintain sections of the A1(M) and A419/A417 trunk road. The non-recourse infrastructure bond issue of £165 million and a guarantee for the servicing of the bond was obtained from a US insurer, AMBAC.

Covenants

The bondholders have no direct voice in the decisions of the borrowing entity. They exercise control through protective covenants specified in the indenture, which limits certain actions of the borrower. Protective covenants can be classified in two categories: negative and positive.

Negative covenants limit or prohibit the action that the borrower may take. For example:

- an upper limit on the debt/equity ratio
- a limitation on the amount of dividend that can be paid
- a limitation on pledging the assets of the company to other lenders
- a limitation on selling or acquiring major assets without prior approval of the bondholders
- a limitation on the borrower merging with another entity.

Positive covenants specify an action that the borrower agrees to take or a condition that the borrower must abide by. For example:

- the borrower agrees to maintain its working capital at a minimum level
- the borrower agrees to furnish periodical financial statements to the bondholders.

Sinking fund arrangement

Bonds can either be repaid entirely at maturity or be repaid before maturity. The repayment takes place through a sinking fund. A sinking fund is an account maintained by the bond trustee for the repayment of bonds. Typically, the borrower makes annual payment to the trustee. Depending on the indenture agreement, the trustee can either purchase bonds from the market or can select bonds randomly using a lottery and purchase them, generally at face value. A sinking fund has two opposing effects on the bondholders:

- It acts like an early warning system, for the lenders, when the borrower is in financial difficulties and unable to meet the sinking fund requirements.
- It is beneficial to the borrowers both when the price of the bond is high as well as when it is low. In the event of a lower market bond price the borrower buys back the bonds at the lower market price, and in the event of higher market bond price the borrower still buys the bonds at the lower face value.

Call provision

Call provision gives the bond issuer the option to retire or repurchase or call the entire bond issue at a predetermined price over a specified period, prior to maturity. Generally bonds are issued with call provisions and these are therefore also known as *callable bonds*. Normally, the call price is above the face value of the bond. The difference between the call price and the face value is called the *call premium*. When the call provisions are not operative during the first few years of the bond's life, usually 5–10 years, it is referred to as a *deferred call*. During this period the bond is said to be *protected*. Bond issuers want call options on bonds for two reasons:

- in the event of a fall in market interest rates, the bond issuer can pay off its existing bonds and issue new ones at a lower interest cost
- it gives a company the flexibility to rearrange its capital structure if market conditions change.

This does not mean that call options work entirely to the borrower's advantage. In fact the more attractive the provisions regarding the call option are to the borrower, the higher is the coupon rate expected by the investors on the bond issue.

Bond ratings

The success of a bond issue depends, *inter alia*, upon its quality. There are many companies that analyse the investment qualities of publicly traded bonds. They publish their findings in the form of bond ratings. The ratings are determined by using various financial parameters of the borrowing agency, the general market conditions in which the borrower operates, the political situation of the country in which the project is located and other sources of finance that have been tied up by the project. Ratings are based, in varying degree, on the following considerations:

- the likelihood of default by the bond issuer on its timely payment of interest and repayment of principal
- the nature and provisions of the obligations
- the protection afforded by the indenture in the event of default, bankruptcy, reorganisation or other arrangements under the law and the laws affecting creditors' rights.

The ratings are normally expressed as a series of letters or a combination of letters and numbers. In certain financial markets, such as the US bond market, the public issue of bonds is not permitted if the bonds have not been rated. Rating is also important because bonds with lower ratings tend to have higher interest costs. The rating agencies keep reviewing the financial performance of the borrower, the general market situation and the political situation in the country of the borrower. Depending on the emerging situation the ratings are revised upwards or downwards. A summary of the common long-term ratings assigned by the two leading bond rating agencies, Moody's Investor Service and Standard & Poors, is given in Table 3.1.

There are some advantages of bond financing over traditional bank lending. Firstly, bond issues can have maturity of up to 30 years, whereas banks usually prefer to lend for a shorter period of time, depending on the type of project. For example, Greenwich Hospital raised £91 million through a 30-year bond issue. It would have been almost impossible for the hospital to secure 30-year funding through the banks. Secondly, bond financing can reach a wider group of investors, and therefore can achieve a lower interest cost margin and longer maturity. Thirdly, bonds are attractive to long-term fixed-income investors because they are backed by the long-term identifiable cash flows of a project.

Bonds offer flexibility, but in some cases the draw down on bonds may not be as previously predicted. This means that promoters could be left with monies that they will have to pay dividends on, even if the money has not been used in the project. In many cases, this money may be reinvested or used as working capital.

The disadvantages of bonds are that in projects with long construction periods the contractors do not need all the money upfront. Also, bond financing does not provide the same degree of flexibility or control of a creditor's interests in the project as does bank lending. Markets can be very unpredictable and borrowings can thus be affected due to shifts in sentiments rather than any reason related to the project.

Table 3.1. Summary of the common long-term bond ratings[*]

Bond rating

Standard & Poors	Moody's Investor Service	Comments
High grade bonds		
AAA	Aaa	Capacity to pay interest and the
AA	Aa	principal is extremely strong
Medium grade bonds		
A	A	Strong capacity to pay interest and
BBB	B	repay the principal, although they are somewhat more susceptible to the adverse effects of changes in circumstances and economic conditions. Both high grade and medium grade bonds are investment quality bonds
Low grade bonds		
BB	Ba	Adequate capacity to pay interest
B	B	and repay the principal. However,
CCC	Caa	adverse economic conditions or
CC	Ca	changing circumstances are more likely to lead to a weakened capacity to pay interest and repay the principal. They are regarded as predominantly speculative bonds. BB and Ba indicate the lowest degree of speculation and CC and Ca the highest
Very low grade bonds		
C	C	The C rating is reserved for income
D	D	bonds on which no interest is being paid. The D rating indicates 'in default', and payment of interest and/or repayment of the principal is in arrears

[*]At times, both rating agencies use adjustments to these ratings. Standard & Poors uses plus and minus signs: A+ is the strongest and A– the weakest. Moody's uses 1, 2 or 3, with 1 being the strongest

Bond financing is being touted as one of the biggest potential forms of funding in Asia, and several projects have been financed using bonds in Australia and China. An example is Hopewell's US $600 million bond issue for the construction of the Guangzhou–

Shenzhen Superhighway. Bank Negara, Malaysia's national bank, has recently relaxed its approval procedures for corporate bond issues by granting a general approval from the Controller of Foreign Exchange for such issues in a move to spur fund-raising activity in the bond market. Approvals will be granted so long as funds raised from the bond issues are not used to finance investments abroad or to refinance offshore borrowings. The bank is aware that a well-developed bond market would remove some of the credit risk from the banking sector, having learned its lesson from the 1997 Asian financial crisis

The launch of the single European currency in 1999 has rapidly increased the number of bond issues denominated in Euros. According to Capital Data BondWare, bonds denominated in Euros accounted for almost 45% of all international bond issuance, almost neck and neck with bonds denominated in US dollars (Merna and Owen, 1998). Other global currencies, including British Sterling, Japanese Yen and Swiss Franc, account for the remaining 10%. The creation of the Euro has reduced the funding costs for companies willing to borrow from the capital markets. Also, it has led to an explosion in mergers and acquisitions in Europe, which has fuelled demand in large-scale financing from the capital markets.

Several projects in the UK have also been financed using bonds. Recent bond-financed projects include (Merna and Owen, 1998):

- the £165 million issue of 25-year bonds for Road Management Consolidated
- the £165 million issue of 24-year bonds for the City–Greenwich–Lewisham Rail Link, an extension of London's Docklands Light Railway
- the £137.4 million issue of 32-year bonds for the rebuilding of South Tees hospital near Middlesborough.

The Mayor of London recently proposed to finance improvements to London Underground by raising funds through the issue of bonds, though his plan has met with disapproval by the Labour government, who prefer to fund improvements to the tube through public–private partnership – uniting private money with government subsidy. The idea of funding improvements to London Underground through bond issue is based on the bond financing used by the Metropolitan Transportation Authority (MTA) in New York, which has successfully improved the conditions of the New York Subway through the recent US $11 billion bond issue. In this project, the management of the subway remains in the public domain.

Other forms of bonds

The bonds discussed so far are known as *plain vanilla bonds*. The other types of bonds are junk bonds, floating-rate bonds, deep discount bonds and revenue bonds.

Junk bonds

Bonds with a Standard & Poors rating of BB and Moody's Investor Service rating of Ba and below are described by the investing community as 'junk bonds'. They are also known as *high-yield* or *low-grade bonds*. Development of the junk bond market during the 1980s was based on the observation that the premium of 5–6 percentage points for junk bonds, compared to safer government bonds, was much higher considering a premium of only 1–2 percentage points required to compensate for the default risk of these bonds. However, the junk bond market collapsed in 1990 due to the high default rate, legal problems, ill-structured bond issues, the bankruptcy of its leading proponent, Drexel Burnham Lambart, and the conviction of its originator, Michael Milkin.

Floating-rate bonds

The coupon rate is tied to some short-term interest rate such as the 90-day Treasury-bill rate or the 6-month LIBOR. The popularity of floating-rate bonds is due to the inflation risk found in the conventional plain vanilla bonds. When inflation is high and interest rates move up, the purchasers of fixed-rate bonds are at a disadvantage. Floating-rate bonds reduce this risk because short-term rates rapidly adjust to the increase in inflation. The adjustment of interest in floating-rate bonds also affects the market price of bonds. Floating-rate bonds normally sell at, or near, par.

Deep discount bonds

No interest is paid. The bond is sold at a discount to the par value. These are also known as *original issue discount bonds*, *pure-discount bonds*, *zero-coupon bonds* or *zeros*. For example, in 1982, PepsiCo issued the first zero-coupon bond. The price on PepsiCo's 30-year bond was around US $60 for each US $1000 face amount of the bonds. The investors' gain comes in the form of the difference between the discounted price and the face value they expect to receive at maturity. These securities appeal to those investors who want to be certain of their long-term return. The locked-in return means that investors know in advance how much money they will receive at maturity, an important consideration for pension funds and other buyers who have fixed future commitments. Conventional bonds do not have this certainty because investors cannot know at the time they purchase the bond what interest rate they will earn when they reinvest their coupon payments.

Income bonds

These are similar to conventional bonds, except that the coupon payments depend on the borrower's income. In particular, coupons are paid to bond holders only when the project's income is sufficient. From the issuer's point of view, income bonds are a cheaper form of debt because they provide the same tax advantages as the conventional bond and the issuer is not at default in the event of financial distress.

Euro bonds

These are mostly US dollar denominated bonds, which are sold in a country other than the country of issue. A US company can issue such a bond in the European market. The funds for the issue are gathered internationally and made available to the borrower. The lending is not influenced or regulated by the national authorities of that currency, and dealers lend directly to the borrower on this market without the bank acting as intermediary. This market was developed by US borrowers in order to avoid the stringent US interest rate regulations. This system can, however, function only if governments allow free movement of currency into and out of the country. Currently, many of the borrowers in this market are public bodies or international organisations. Eurobonds are usually bearer bonds rather than registered, which pay interest at fixed or variable rates on the presentation of a coupon. Handing over the bond to a new owner transfers ownership. Only listed borrowing entities are allowed to access this market. As in the case of normal bonds a trustee represents the interests of the borrower.

Summary

Bonds issued to finance specific infrastructure projects can often provide income tax breaks to the investor equivalent to a percentage of the value of the bonds purchased or on the earnings of the bondholder. Even with equal coupon and interest rates this financial instrument should earn small investors better returns in real terms when compared with ordinary savings accounts. The type of bond issued should be compatible with the revenue generated by a project. The ordinary, plain vanilla bond issue should be used for projects with ongoing revenue generation, as coupon payments are normally made on a biannual or annual basis. For infrastructure projects where revenue is generated after an extended period of time or in one lump sum, bonds of zero coupon bonds may be used. High-risk projects should not raise finance through plain vanilla bonds, but should instead use zero coupon, revenue or high-risk (junk) bonds.

Private placement of bonds

The private placement of bonds is similar to obtaining a term loan. The bond is placed with a select number of investors, normally banks, financial institutions, pension funds, housing societies and other institutional investors. Raising finance through this mechanism does not require registration with securities exchange authorities such as the SEC in the USA. This saves money. Advantages of this type of bond placement include:

- the maturity of privately placed bonds is generally longer than that of term loans
- the cost of distributing bonds in the private market is lower
- it is easier to renegotiate the terms and conditions of the bond in the event of default.

Disadvantages include:

- the interest rate on privately placed bonds is higher than for the public issue of bonds.
- there are likely to be more restrictive covenants.

Preference shares

These are the shares that possess priority rights over ordinary shares. Preference shares entitle the holder a preferential right over lower ranked ordinary shares, both in terms of dividend and return on capital, in the event of liquidation. Normally the preference share holders have the right to a fixed annual dividend, the right to receive repayment of any amount paid up on the preference shares on a winding up, and restricted voting rights. The board of directors of the issuing organisation may decide not to pay the dividend on preferred shares, and this decision may have nothing to do with the current income of the issuer organisation. The dividend payable on the preference shares is either cumulative or non-cumulative. If cumulative dividends are not paid in a particular year they are carried forward. Usually both the cumulative preferred dividend and the current preferred dividend must be paid before ordinary shareholders can receive anything. Unpaid dividends are not treated as debt. The issuer organisation may decide to defer the payment of dividend on preferred shares indefinitely. However, if they do so, the ordinary shareholders also do not receive anything. It is argued that preferred shares are in fact debt in disguise. The preferred shareholders receive only a stated dividend, and a stated value in the event of liquidation of the issuing organisation. However, unlike interest on debt, dividend on preferred shares is not deductible before determining the taxable income of the borrower.

Warrants

Warrants are securities that give the right, but not the obligation, to buy ordinary shares from a company at a fixed price during a given period of time. Each warrant specifies the number of shares that the holder of warrants can buy, the price at which he can buy and the expiration date. The fixed price and given period of time are referred to as the *exercise price* and the *exercise period*, respectively. Some warrants are perpetual as they never expire.

A significant proportion of private placement bonds and a smaller proportion of public issue bonds are sold with warrants. In addition, warrants are sometimes given as compensation to underwriters. They can also be attached with new issues of ordinary shares and preferred shares. Since warrants are usually issued in combination with privately placed bonds they are frequently referred to as *equity kickers*. Warrants are normally attached to the bonds when issued. It is possible to detach them and trade them separately. In such a case these are known as *detachable warrants*.

When the holder exercises the warrant option, the issuer of the warrant has to issue new shares in return for the exercise price specified in the contract. Warrants have no voting rights, do not pay interest, and offer no claim on the assets of the issuer.

The value of a warrant depends on its expiration date, exercise price and the underlying share price. Other things being equal one may say that the higher the exercise price the lower the value of a warrant, and the higher the price of underlying shares the more valuable the warrant. Also, the value of a warrant must be at least as great as the value of another warrant with a shorter term of expiration.

Convertible bonds

A convertible bond gives the holder of the bond the right to exchange it for a given number of shares up to and including the maturity date of the bond. A convertible bond is similar to a bond with a warrant. The most important difference is that, whereas a bond with a warrant can be separated into distinct securities, a convertible bond cannot. The difference between a convertible preferred share and convertible bond is that the former has an infinite maturity date and the latter has a finite one. The number of shares received for each bond is called the *conversion ratio*. At the time of issue of convertible bonds the conversion ratio is specified, such as £10 of bonds may be converted to eight £1 shares. Convertibles are usually subordinated and unsecured.

Both warrants and convertibles are issued by high growth and better than average financial leverage projects. Projects prefer to issue convertible debt because the interest payable on this type of debt is lower than the straight debt. Investors accept low returns on their investment in the hope of potential gains from conversion. The hope, however, becomes true only if the project performs well and its shares sell at a price higher than the conversion price quoted in the convertible bond.

Other financing instruments

Depository receipts

A depository receipt is a negotiable certificate representing a project's equity or debt. A depository bank appointed by the project creates depository receipts. New as well as old shares of the project can be made available to an appointed depository's local custodian bank, which then instructs the depository bank to issue depository receipts. These depository receipts are thereafter issued either publicly or through private placement. Depository receipts may be traded freely, just like any other security. Depository receipts in US form are known as *American depository receipts* (ADR) and in global form they are known as *global depository receipts* (GDR).

A project can access the US market and other markets outside the USA either through a private placement of depository receipts or public offer. With a public offer a detailed procedure of SEC registration is required. However, through a private placement of depository receipts, the project can raise capital by placing depository receipts with large institutional investors in the USA, avoiding SEC registration, and also to non-US investors.

During 1997, of the US $5642 million raised through private placement of GDRs the maximum amount was raised by India (19.3%), followed by Taiwan (17.0%), Hungary (10.4%), Poland (7.2%), Brazil (7.1%) and Korea (7.1%).

Lease finance

Lease finance is a contractual arrangement between a lessee and a lessor. The agreement establishes that the lessee has the right to use an asset and in return must make periodic payment to the lessor, the owner of the asset. The lessor could be either the manufacturer of the asset or an independent leasing company. If the lessor is an independent leasing company then it must buy the asset from a manufacturer and make it available to the lessee for the lease to be effective.

Leases are principally of two types: operating lease and financial lease.

Operating lease

An operating lease can be used for assets that are required occasionally. The important features are:

- Operating leases are usually not fully amortised. This means that the payments required under the terms of the lease are not sufficient to recover the full cost of the asset for the lessor. This is because the term or life of the operating lease is usually less than the economic life of the asset. The lessor therefore must either renew the lease or sell the asset for its residual value.
- Operating leases usually require the lessor to maintain and insure the leased assets. There is generally a cancellation option in this type of lease which gives the lessee the right to cancel the lease contract before the expiration date.

Financial lease

In a financial lease the potential user identifies an asset in which it wishes to invest, negotiates a price and delivery, and then seeks a supplier of finance to buy it. The asset is bought by the lessor and made available to the lessee for use. The important characteristics of financial leases are:

- financial leases do not provide for maintenance or service by the lessor
- financial leases are fully amortised
- the lessee usually has the right to renew the lease on expiration
- financial leases generally cannot be cancelled.

The two other types of financial leases are sale and lease back and leveraged leases.

Sale and lease back. In this type of financial lease the owner of an asset sells the asset to a lease company and immediately leases it back. This helps the lessee generate cash from the sale of the asset and at the same time keep using the asset by making periodic lease payments.

Leveraged lease. A leveraged lease is a three-sided arrangement among the lessee, the lessor and the lender. The lessor buys the asset and provides it for use by the lessee. However, the lessor puts in no more than 40–50% of the purchase price and the remaining financing is obtained from a lender. The lender typically provides a non-recourse loan and receives interest from the lessor. The lender has the first lien on the asset, and in case of default by the lessor the lease payment is made directly to the lender.

Venture capital

Venture capital consists of funds invested in small, newly established enterprises, normally in high growth markets, which promise exceptional future profit levels. The investment is normally made in the form of equity capital. However, some venture capital funds invest in a mixture of debt and equity capital. Although venture capital is thought to be providing capital only for new ventures, its role is much wider. The types of investment that are being financed under venture capital can be put under five categories: start-up, early stage development capital, management buy-out (MBO), expansion and others.

Aid

Aid refers to a direct gift of money from one government or a multi-lateral such as the World Bank or government agency to another government. Grant aid is usually intended to help less developed countries meet social or community welfare objectives. It is seldom free from obligations on the recipient government. It can have direct strings attached to the project or may influence the way a project is managed. Other forms of aid are in the form of subsidised credit export or an aid plus credit package. These are best considered as debt financing. Aid is basically of two forms:

- *project aid* is specific, highly structured and formalised in accordance with the donor institution's project appraisal procedures
- *programme aid* intends to finance imports in return for sectoral policy reforms; it may involve individual projects.

Aid finance from one government to another government is known as *bilateral aid*. It is usually, but not always, tied to supplies from the donor country. When aid is from a multi-lateral institution it is known as *multi-lateral aid*. Use of funds in the case of multi-lateral aid is not tied to any particular supplier.

Short-term financing instruments

Projects require short-term debt of two kinds. Firstly, it is required for working capital. These funds are needed once the project has been commissioned, and there is a time lag between payments to be made for the purchase of raw materials, components and equipment for the operation of the project and receipt from the sale of the product. Secondly, short-term debt is required as bridging finance to meet temporary deficits in cash balance when there is a known source of funds that can be fully relied upon to liquidate the bridging loan. This may be due to various reasons, such as the project may have negotiated a long-term loan but the formalities of documentation and fulfilling of various covenant conditions attached to the loan is taking longer than anticipated and the project is in need of immediate resources.

Alternatively, the project may have raised equity capital through public offer but registration formalities with the local stock exchange authorities is taking longer or the project is waiting to launch the issue when the stock market is in the right condition. In such situations the project raises bridging loans to fund the immediate requirement. The bridging loan is liquidated from the proceeds of the main loan or the equity capital raised.

The options available to a project for working capital requirements during the operation phase, other than the internal resources, can be broadly classified into three categories:

* unsecured bank borrowing
* secured borrowing
* other sources.

Unsecured bank borrowing

The most common method of financing a temporary cash deficit is to arrange a short-term unsecured bank loan. This is either in the form of a non-committed or a committed line of credit.

A *non-committed line of credit* is an informal arrangement that allows the project to borrow up to a previously specified limit without going through the normal loan documentation requirements. The interest rate on the line of credit is usually at the bank's prime lending rate plus an additional charge. Generally a compensating balance (about 2–5%) is required to be kept at the bank.

A *committed line of credit* has a formal legal arrangement and documentation requirement. It also involves a commitment fee. The interest rate is often linked to LIBOR or to the bank's cost of funds, rather than to the prime rate.

Secured loans

Banks and financial institutions often require security for a loan. The security for a short-term loan usually consists of accounts receivable or inventory. In accounts receivable the lender has a lien on the receivable of the project and also recourse to the borrower's assets.

Other sources

There are a variety of other mechanisms to secure short-term financing. The most important are commercial paper and banker's acceptance.

Commercial paper

Commercial paper consists of short-term notes issued by large and highly rated projects. Typically these notes are of short maturity, ranging up to 270 days. The issue is normally backed by a special line of credit. Therefore the interest rate on commercial papers is below the prime lending rate.

Banker's acceptance

A banker's acceptance is an agreement by a bank to pay a sum of money. These arrangements typically arise when a seller, such as a supplier of raw materials or components, sends a bill or draft to a customer. The customer's bank accepts this bill and notes the acceptance on it, which makes it an obligation of the bank. In this way a project that is buying something from a supplier arranges for the bank to pay the outstanding bill. The bank charges a fee for this facility.

Bibliography

Lessons of Experience No. 4: Financing Private Infrastructure. International Finance Corporation (IFC), Washington, DC, 1994.

Merna, T. and Dubey, R. *Financial Engineering in the Procurement of Projects.* Asia Law and Practice, Hong Kong, 1998.

Merna, T. and Owen, G. *Understanding the Private Finance Initiative.* Asia Law and Practice, Hong Kong, 1998.

CHAPTER FOUR

Financial engineering

Introduction

Just as engineers use special tools and instruments to achieve engineering perfection, financial engineers use specialised financial instruments and tools to improve financial performance. The International Association of Financial Engineers (IAFE) describes the term 'financial engineering' as:

> The development and creative application of financial technology to solve financial problems and exploit financial opportunities.

Galitz (1995) defines the concept of financial engineering as:

> The use of financial instruments to restructure an existing financial profile into one having more desirable properties.

Financial engineering techniques are increasingly being applied widely, such as in the modelling and forecasting of financial markets, the development of derivative instruments and securities, hedging and financial risk management, asset allocation, and investment management and asset or liability management.

Financial engineering tools

The tools used by financial engineers comprise the new financial instruments created during the last two to three decades, these being forwards, futures, swaps and options. Financial engineers are now combining the basic tools in different ways to build more complex systems to meet specific requirements of their clients. The term *financial architecture* is sometimes used to cover all the parties involved in financing a project and all the instruments used. Merna and Dubey (1998) have described the basic instruments and some of their variants such as forward rate, forward exchange rate, forward interest rate and futures.

Forward rate

The forward rate is the price the market sets for an instrument traded today, but where the resulting transaction is executed at some date in the future. The most common forward rates used in the financial world are forward exchange rates and forward interest rates.

Forward exchange rate

The forward exchange rate can be described best using an example. Suppose a US based firm has to pay DM 1 100 000 to one of its suppliers or one of the agencies from which it has borrowed, in 1 year's time. Since the firm is based in the USA it will have to buy Deutschmarks with its dollar income. In this era of floating exchange rates, suppose the firm wants to fix its liability now and approaches a bank and requests a quote for Deutschmarks against the dollar, for delivery 1 year from now. The question then arises, what is the fair dollar–Deutschmark exchange rate for 1 year in the future? When exchange rates are fluctuating with every deal that is made in the market, it is difficult to predict exchange rates in the distant future. Fortunately, it is not necessary for the bank to approach a soothsayer for a prediction. The bank can price a forward exchange (FX) deal using the principle of risk-free arbitrage. The principle of risk-free arbitrage works in the following way.

The bank knows the spot dollar–Deutschmark exchange rate. The *spot rate* is the rate at which any two currencies sell in the current market. Suppose the spot rate is US $1 = DM 1.6, and the current 1-year dollar interest rate and the 1-year Deutschmark interest rate are 6% and 10%, respectively. If the bank agrees to sell DM 1 100 000 to its client at the end of 1 year, it must have this amount of money at that particular point in time. The bank can ensure the availability of this amount by lending a sufficient amount of Deutschmark today for 1 year, so that it receives the principal plus the interest amount exactly equal to DM 1 100 000 1 year from now. At the current Deutschmark interest rate of 10%, the bank would need to lend DM 1 000 000.

The question that arises is from where does the bank get the Deutschmark? The bank can purchase Deutschmark from the spot market at the spot rate of US $1 = DM 1.6. In order to raise DM 1 000 000 the bank will have to spend US $625 000 at this rate.

Even now the bank is left with a gap of US $625 000. The bank can fund this amount by borrowing from the market at the prevailing interest rate of 6%. But at the end of 1 year the bank has to repay its lender the principal amount of US $625 000 plus the interest. This comes to US $662 500 at the interest rate of 6%. The bank has to pay the dollar liability 1 year from now and its client is also willing to provide dollars 1 year from now in exchange for Deutschmark. The bank can match its receipt and payment, 1 year from now, by quoting a price of US $662 500 for the DM 1 100 000. In this way the full circle has been completed. The bank has been able to cover all its positions, its receipts and payments match at the end of 1 year, and it has been able to service its client by quoting a price for Deutschmark against the dollar at a future date without involving any money of its own. In this process the dollar–Deutschmark forward exchange rate is US $662 500 = DM 1 100 000, i.e. US $1 = DM 1.66 can be derived.

Since the entire structure of pricing a forward exchange is based on the market today, there is no need for the bank to have any opinion about what the spot rate will be in the future. This is the essence of risk-free arbitrage pricing.

Forward interest rate

The forward interest rate can best be understood with the help of the following example. Suppose a firm needs to take a loan 6 months from now to meet some of its liability. Since the interest rates keep changing it wants to protect itself against the exposure of higher borrowing cost and approaches its bank for a quotation for taking a loan 6 months from now at a fixed interest rate. In this case both the draw down of the loan and the repayment of the loan is in the future. In view of these features, such loans are also known as *forward–forward loans*.

Since the bank does not wish to take any risk it would like to fix its financing cost. In the normal financing market there is either 6-month lending or 12-month lending and, therefore, there is either a 6 months interest rate or 12 months interest. But the bank, in this case, has been requested to quote a rate for 6-month lending where funds will be drawn in the future, not now. The bank can again resolve this problem, by using the principle of risk-free arbitrage pricing, to quote a fair price.

To fix the cost for the 6-month period loan starting in 6 months, the bank will have to borrow now for 12 months at the prevailing interest rate for 12-month borrowing. Now the question arises of how much the bank should borrow. And what will it do with the funds for the initial 6 months since the client needs the funds 6 months from now? We can answer the second question first. The bank will lend this amount for 6 months at the 6-month interest rate. The proceeds of the 6-month loan, principal plus interest, should be sufficient to meet the client's requirement. Since the 6-month interest rate is known, one can determine the amount of money the bank should lend. This is the amount that the bank should borrow. This has answered our first question. Now the bank should lend to the client at an interest rate such that it is able to meet its requirement to service the 12-month loan it has taken. This is the interest rate that the bank should quote to the client.

Briefly, the bank will borrow for 12 months at the prevailing 12-month rate, and for the first 6 months it will lend at the prevailing 6-month rate. For the remaining 6 months it will lend at such a rate to its client so that the bank receives sufficient funds to meet its liability on the 12-month loan it has taken.

As in the case of the forward exchange rate, in this case the bank is also in a position to quote a price for a 6-month loan starting 6 months from now, entirely with the help of current market rates of interest. It does not have to form any opinion about the future interest rate.

Forward rate agreement

Galitz (1995) defines a the forward rate agreement (FRA) as:

> ...an agreement between two parties motivated by the wish either to hedge against or to speculate on a movement in future interest rates.

In essence, an FRA is a forward–forward loan at a fixed interest rate, but without the actual lending commitment. If flows of the principal amount are removed from the transaction, it takes this instrument off the balance sheet. Banks will, however, still be required to allocate some capital to protect their position.

An FRA is an agreement between two parties to modify their interest rate exposures. The two parties are known as the *buyer of the FRA* and the *seller of the FRA*. The seller of an FRA agrees to lend a particular or notional sum of money to the buyer. The notional sum is of a specified amount in a specified currency and will be drawn down on a particular date in the future, and will last for a specified period. Most importantly the notional loan will be made at a fixed rate of interest, this rate being agreed when the FRA deal is struck.

The buyer of an FRA is therefore a notional borrower, and is protected against a rise in interest rates, though he must pay if rates fall. The buyer may have a real borrowing requirement, and is using the FRA as a hedge. Alternatively, the buyer may have no underlying interest rate exposure, but may be using the FRA simply to speculate on a rise in the interest rate.

The seller of an FRA is a notional lender, and fixes the rate for lending or investing. The FRA seller is therefore protected against a fall in interest rates, but must pay if interest rates rise. A seller may be an investor who would really suffer if rates fell, but could also be someone with no underlying position who just wants to profit from a fall in rates.

Although one or both parties may have borrowing or investing requirements these are handled separately. No lending or borrowing actually takes place under the FRA itself. The FRA simply provides protection against a movement in interest rates. This is achieved through a cash payment, the *settlement sum*, which compensates each party for any difference between the rate of interest originally agreed and that prevailing when the FRA eventually matures.

For example, a company needs to borrow US $1 million in 3 months time for a 6-month period. Let us assume that the company can borrow at the London Interbank Offer Rate (LIBOR) flat, which is, say, 6%. The borrower, however, expects the interest rate to rise in the next 3 months. If the borrower does nothing he could face a much higher interest rate when the loan is drawn down in 3 months time. To protect his interests the borrower could buy an FRA today to cover the 6-month period starting 3 months from now. This would be known in the market as a *3-against-9 month FRA*, or simply a 3X9 FRA. A bank may quote a rate of 6.25% for such an FRA, and this would enable the borrower to lock into a borrowing rate of 6.25%.

Now if the interest rate rises during the 3-month period to 7% then the borrower will have to borrow from the market and pay the going rate, namely the 7%. But the FRA will come to his rescue at the time of repayment. The borrower would receive an equivalent amount from the seller of the FRA to compensate him for the difference in the interest rate of 0.75% (7% less 6.25%). The settlement sum effectively offsets the borrower for the higher borrowing cost. We have seen that buying an FRA has not guaranteed the interest rate on the specific financing facility used by the borrower, but he has been able to secure financing at a fixed interest rate, which is of paramount importance to project sponsors.

All FRA deals in the UK have standardised documentation drawn up in 1985 by the Forward Rate Agreement British Bankers Association (FRABBA). The standard documentation defines a number of important terms:

- *contract amount* – the principal sum (notional) lent or borrowed
- *contract currency* – the currency in which the contract amount is denominated
- *dealing date* – the date when the FRA deal is struck
- *settlement date* – the date when the notional loan or deposit commences
- *fixing date* – the date when the reference rate is determined
- *maturity date* – the date when the notional loan or deposit matures
- *contract period* – the number of days between the settlement and maturity dates

- *contract rate* – the fixed interest rate agreed under the FRA
- *reference rate* – the marked base rate used on the fixing date to determine the settlement sum
- *settlement sum* – the amount paid by one party to the other on the settlement date, based on the difference between the contract and reference rates.

Financial futures

Long before the existence of organised grain and commodity markets, farmers would bring their harvested crops to major population centres in search of buyers. There were no storage facilities, and many times the harvest would rot before buyers were found. Many farmers would also bring their crops to market at the same time, which resulted in the price of crops or commodities going down. If, however, there were a shortage of crops and commodities then prices would rise sharply. This was because of a mismatch between supply and demand. There was no organised or central marketplace where competitive bidding could take place.

Initially, the first organised and central marketplaces were created to provide spot prices for immediate delivery. Shortly thereafter, forward contracts were established. These 'forwards' were forerunners to the present day futures contract.

The Chicago Board of Trade (CBOT) was the first to offer futures contracts on agricultural products, in the middle of the nineteenth century. The need for, and therefore organised development of, the financial futures market started in the early 1970s due to increased volatility of interest rates and exchange rates after the breakdown of the Bretton Woods system of exchange rates. As in the case of commodity futures, the CBOT was the first centre where organised futures trading started in 1975. Gradually it expanded to other financial centres: London, 1982 (London International Financial Futures Exchange (LIFFE)); Paris, 1986 (Marché A Terme des Instruments Financiers (MATIF)); Switzerland, 1988 (Swiss Options & Financial Futures Exchange (SOFFEX)); Dublin, 1989 (Irish Futures and Options Exchange (IFOX)); Germany, 1990 (Deutsche Termin Borse (DTB)); Austria, 1991 (Oesterreichische Termin und Optionborse (OTOB)); and Italy, 1992 (Multilateral Investment Fund (MIF)).

The futures market in currency and interest rates gradually spread to interest-bearing assets and stock indices. The futures contracts can be classified into the following four categories:

- short-term interest rate futures, such as the Eurodollar, Euroyen, 3-month Sterling and Euromark

- bond futures, such as the US T-bond, T-notes, French Government Bond and British Gilt
- stock index futures, such as the Standard & Poors 500, Nikkei 225 and FTSE 100
- currency futures, such as the Deutschmark against the US dollar.

Galitz (1995) defines a futures contract as:

> ... [a] legally binding agreement to take or make delivery of a given quantity and quality of a commodity at an agreed price on a specific date or dates in the future.

Dixon and Holmes (1992) define futures contract as:

> An agreement to buy or sell a standard quantity of a particular commodity or financial instrument at a future date for a price which is agreed at the time the contract is drawn up. Just like any other contract a future contract involves an obligation on the part of both the buyer and the seller to fulfil the conditions of the contract.

The standardisation of futures contracts is perhaps the most important factor that distinguishes futures contracts from forward contracts. Each futures contract is for a standard specified quantity. For example, each LIFFE 3-month, dollar interest-rate futures contract has a unit of trading of US $1 000 000. The contract not only specifies the quantity of the instrument in the contract, but also the quality in terms of its coupon rate and maturity.

The great advantage of the standardised contract is that potential buyers and sellers of the contract can easily find out the details of the contract, whereas with forward contracts the market participants will have to acquire information about each individual contract. Since futures contracts are not designed for specific parties, market participants can easily close out their position by taking an offsetting position in the market. Standardisation increases the marketability of the contracts, and therefore the liquidity within the market.

Buying and selling in the futures market

There is a fundamental difference between the cash market and the futures market with regard to the act of selling and purchasing. In the cash market, market rules normally impose restrictions on selling short, that is selling something that one does not possess, such as a product or a facility. In the futures market there is no such distinction. Buying and selling of contracts are completely symmetrical transactions. This is entirely appropriate for a futures transaction because a futures contract binds the parties to a transaction that will take place in the future. A party not possessing the underlying asset can very well acquire the asset or reverse the futures contract before its maturity.

Basic trading strategies in futures markets

- *Buying (going long) if price is expected to increase.* An investor expecting the price of a futures contract to increase may decide to go long, i.e. to purchase a futures contract. If the price of the contract rises the investor will profit. If the price of the contract declines, the investor will lose money.
- *Selling (going short) if price is expected to decrease.* An investor who believes that the price of a futures contract will decline sells a futures contract. The mechanics of selling short are that, first, a futures contract is sold, and then the profit is realised by buying an offsetting contract at a lower price. If the price of the contract declines the investor makes a profit. If the contract price increases the investor incurs losses.

Whether trading of a futures contract takes place in the trading pit or on the screen of a clearinghouse, the clearinghouse is interposed between each buyer and seller. This means that the obligation of the buyer and the seller is not to each other but to the clearinghouse. Once the deal between the buyer and the seller has been struck, the clearinghouse takes over the role of other parties to the transaction. Thus the clearinghouse is the seller of the contract to the buyer and the buyer of the contract from the seller. By coming between the buyer and the seller the clearinghouse removes the counter party risk. The identity of the counter parties is therefore not of much significance. The financial resources of the clearinghouse ensure that every transaction that is carried out on the exchange is guaranteed. It is the large volume of transactions that makes the clearinghouse settlement process economically viable. Removing the default risk means that trading can proceed quickly without pausing to check on individual credit limits. The only two variables on which the trading counter parties have to agree are the size and price of the transaction. The clearing mechanism has another important advantage that, no matter how many times one enters the market to buy or sell, or no matter when the deals are struck, there is always one ultimate counter party. By allowing transactions to be closed out by opposite transactions the customer can end up with no position. This feature of the clearinghouse adds to the liquidity of the market. This is in sharp contrast to the cash market transaction, where if a customer buys an FRA from a bank and sells an FRA to another bank, then in fact there are two FRAs on the customer's books.

The clearinghouse takes all the counter party risks of default. However, the potential for loss is enormous, considering the volume of futures contracts traded and the underlying positions they represent. The clearinghouse protects itself from this risk through the system of *margin* requirement. The specification of every futures contract defines the level of margin that must be deposited by the member with the clearinghouse. For example, if the margin required for positions in the Eurodollar futures traded on the CME is US $500 per contract, then if a member holds a long position of 100 contracts he has to deposit a cash margin of US $50 000. Both the buyer and the seller of the contract have to deposit the same amount of margin. This margin is, however, in no way a down payment for buying or selling the contract. It is like a guarantee of the market participants for fulfilling the contract. The margin acts as a performance bond (Galitz, 1995). When the members eventually close out a position, the margin is refunded.

The initial margin is set by the clearinghouse, depending on the volatility of the underlying instrument. Normally this is about 5% of the face value of the contract. However, since many contracts have very long maturity and if margin is deposited just once in the beginning and refunded when the position is closed at maturity, then the potential loss could be sizeable. To make the system work and to protect the clearinghouse from day to day movement in the price of the contract all open future positions are marked to the market at the close of each trading day. This involves the use of *variation margin*, whereby each day's gains are added to the margin account and losses subtracted from that account.

The initial margin must be maintained and, as a result, additional funds will be required when losses are made. All losses are collected each day as the position is marked to the market. In the case where the contract is held until delivery the buyer must pay the full value of the contract to the seller.

Financial futures markets were established with the prime objective of enabling companies and individuals to insure against the possible adverse effects of changes in the interest rates and exchange rates. Thus the main function of the futures market is to help in the reduction of risk, or *hedging*. However, when an individual or a company reduces risk by hedging, the risk is not eliminated but transferred to the counter party. This counter party may be another hedger with opposite requirements, or may be a speculator or trader. Speculators or traders are active in the market not with an objective to reduce risk but to take the risk in an expectation of profits from their activity. They trade on the basis of their expectation about the future changes in the prices.

Futures markets are attractive to traders for two reasons:

- The small margin requirements enable the traders to deal in contracts worth many times the amount of money that they need to commit. This means that futures contracts are highly geared.
- The prices in futures markets are relatively volatile and movements in the basis (the difference between the price of a future contract and the price of the instrument in the spot market) are sufficiently large for potentially high rewards to result.

Advantages and disadvantages of futures

There are several advantages and disadvantages of futures markets. The advantages are:

- *Liquidity.* The standardisation of contracts and the efficiency of trading serve to encourage tremendous liquidity. In a number of cases, the liquidity in the futures market exceeds that in the underlying cash market.
- *Clearing.* The clearing mechanism removes individual counter party risk and allows for the easy reversal of existing futures positions.
- *Margining.* The margining system offers holders of the futures contracts the ability to control large positions in the underlying financial commodity with the minimum of capital. Hedgers can reduce their risk exposure cheaply, and without the physical purchase or sale of underlying instruments. Speculators can exploit their views of market movements without committing vast cash resources or tying up credit lines.
- *Transaction cost.* Futures exchanges aim to keep the trading costs as low as possible. Normally the cost of executing a futures contract is a fraction of the equivalent transaction in the cash market.

The disadvantages are:

- *Inflexibility.* Futures markets require that the specification of each futures contract be rigidly standardised. The cash market allows every aspect of a transaction to be negotiated individually.
- *Liquidity.* While liquidity may be very high for the futures contract with the nearest delivery date, there may be limited liquidity in the back contract for certain futures.

- *Margining*. Some people find that managing the margin account, and the daily cash flow, is a considerable administrative burden.

In summary, standardised contracts, organised exchanges and clearing houses and low margin requirements in futures trading not only help in reducing the risk of certain types of trade, but also help in stabilising the price of the traded assets in the spot market. The primary purpose of futures markets is to provide an efficient and effective mechanism to manage price risk. By buying or selling futures contracts, these contracts establish a price level for today, for items to be delivered later. Individuals and businesses seek to achieve insurance against adverse price changes. This is achieved though buying or selling futures contracts with a price level established today for items to be delivered later.

Differences between a forward contract and a futures contract

A futures contract can be distinguished from a forward contract in the following ways. Firstly, futures contracts always trade on an organised exchange. Secondly, futures contracts have standardised terms. With a futures contract, the quality, quantity and delivery date are predetermined.

Thirdly, futures exchanges use clearinghouses to guarantee that the terms of the futures contract are fulfilled. The clearinghouse is the actual buyer of the contract from the short seller, and the clearinghouse is the actual seller of the long contract. If either party defaults on the contract the clearinghouse steps in and becomes the seller or buyer at the last resort. The clearinghouse guarantees that the contract will be fulfilled. Neither party needs to trust the other party.

Fourthly, margins and daily settlement are required with futures trading. These are other safeguards in the futures market. Each customer must put up a good faith deposit. The amount of this margin varies from exchange to exchange and broker to broker. However, no broker may margin a contract for less than the exchange minimum. Each trading day every futures contract is assessed for liquidity. If the margin drops below a certain level the trader must deposit additional margin. This is called the *maintenance margin.*

Fifthly, futures positions can easily be closed. The trader has the option of taking physical delivery, placing an offsetting trade and arranging an exchange-for-physical transaction. The futures exchange makes executing a contract relatively easy. Finally, forward contract markets are self-regulating and futures markets are regulated by certain agencies dedicated to this responsibility.

Swaps

Financial engineers primarily classify swaps into two categories: interest rate swaps and cross-currency swaps.

Galitz (1995) defines an *interest rate swap* as:

> ... an agreement between two parties to exchange streams of cash flows denominated in the same currency but calculated on different bases.

For example, two parties may enter into an arrangement whereby one party agrees to pay a fixed annual interest of 12% on a notional amount of 100 million in return for the prevailing Sterling pound 6 month LIBOR rate, from the other party. The fixed ratepayer would benefit if the LIBOR rate is higher than 12% in a given period and may lose if LIBOR were lower than 12%. An interest rate swap is, therefore, similar to an FRA but operates over multiple periods.

In a *standard interest rate swap*, also known as *plain vanilla swap*, the two parties who contract to make periodic interest payments to each other are known as the *counter parties*. The payment is normally made on a predetermined set of dates in the future, based on a predetermined principal amount called the *notional principal amount*, denominated in the same currency. In such a swap the party who agrees to pay the fixed rate of interest for the life of the contract is known as the *fixed ratepayer*. The other party who has agreed to pay an interest rate with reference to some floating rate such as the LIBOR or Treasury Bill Rate or Prime Rate, known as the *reference rate*, is known as the *floating rate payer*. There is no exchange of principal; only an exchange of interest takes place. The following example explains how a standard interest rate swap arrangement operates.

Suppose there are two companies. One is AAA rated and the other is BBB rated. A company with a higher rating will be able to procure funds at a lower rate as compared to a company with a lower rating. Suppose the market is offering funds to the AAA rated company at LIBOR + 10 BP in floating interest rate and 11% in fixed interest rate and offers being made to BBB rated company are LIBOR + 50 BP and 12%, respectively. Now, suppose the AAA rated company prefers to borrow floating, and the BBB company preferred to borrow fixed due to various considerations like the existing financial profile of the companies or the decision of the board of directors. By itself, the AAA rated company would borrow at LIBOR + 10 BP and the BBB company would borrow at 12%.

We can now see how by entering into an interest rate swap arrangement both the companies can benefit and at the same time achieve the objective of having their interest rate liabilities in fixed or floating rate terms as decided by the company. Let us consider that the AAA company borrows at a fixed interest rate of 11%. Suppose that the BBB company borrows at LIBOR + 50 BP. Suppose that they enter into an interest rate swap in which the AAA company agrees to pay to the BBB company periodic payments based on the LIBOR in return for receiving periodic payments fixed at 11.20%. For this to work, both borrowing and the interest swap must all be for the same principal amount. For the AAA company the net result of borrowing fixed at 11%, receiving fixed payments at 11.20% and paying the LIBOR, is to borrow at LIBOR + 20 BP. Similarly, the BBB company ends up borrowing at a fixed rate of 11.70%. Both borrowers have managed to borrow using their preferred method of funding, but each at a cost of 30 BP lower than they could otherwise have achieved. In the structuring of such a deal there has been a net gain of 60 BP, which has been equally shared between the two parties.

Other forms of interest rate swaps are detailed below.

Accreting, amortising and roller coaster swaps

In these types of swaps the principal starts off small and then increases over time. In the case of an amortising swap, the principal reduces in successive periods. If the principal increases in some periods and reduces in others, the swap is described as a roller coaster swap. The accreting swap will be attractive for a construction project firm where the amount being borrowed gradually increases during the life time of the project. The amortising swap will be more suitable for hedging a bond issue which features sinking fund payments. For projects where the amount borrowed increases initially and then reduces as stage payments are made, a roller coaster swap will be more suitable. In either case it is not necessary for the underlying principal to follow a regular pattern; the only requirement is that the notional principal for each swap period be defined at the inception of the swap.

Basis swap

In a basis swap both parties may have floating liability but the floats are on different bases. One loan may be linked to the LIBOR and the other to Treasury Rates. Another variant could be that one loan is on a 6-month LIBOR basis and the other is on a 1-month LIBOR basis. Through a basis swap the base can be changed to meet the specific requirements of the firm.

Forward-start swap

In a forward-start swap the effective date is deferred to some future date. A swap counter party may wish to fix the effective cost of borrowing for floating rate financing to be arranged some time in the future. For example, a company may have just won a project mandate, and now be committed to raise finance that will be drawn down on some future date. If the company waits then it runs the risk of a rise in interest rates.

Zero-coupon and back-set swaps

These types of swaps replace the stream of fixed payments with a single payment, either at the beginning or, more usually, when the swap matures. In a back-set swap the setting date is just before the end of the accrual period, not just before the beginning. The floating rate is therefore set in arrears rather than in advance. Such swaps would be attractive to a counter party who holds a view that the interest rate would evolve differently from market expectations.

Cross-currency swap

Galitz (1995) defines a cross-currency swap as:

> ... an agreement between two parties to exchange streams of cash flows denominated in different currencies calculated on similar or different bases.

The cross-currency swap is different from the interest rate swap in the sense that the two streams of cash flows are denominated in two different currencies. For example, one party could agree to pay annual interest fixed at 10% per annum on a notional principal of DM 10 million, while receiving annual floating rate interest based on a 6-month LIBOR on a notional principal of US $6.25 million. The interest rates could both be fixed, both be floating or one fixed and the other floating.

The basic features of a cross-currency swap are:

- it involves two currencies at the two legs
- there is always an exchange of principal at maturity
- there is optionally an exchange of principal on the effective date (i.e. the date from which the deal becomes effective and interest starts to accrue)
- the interest rates could be fixed or floating, or one fixed and one floating.

The following example explains how a cross-currency swap operates. Suppose a US-based company needs Deutschmarks to fund a construction project in Germany. It can issue fixed-rate bonds in

Deutschmarks, or in dollars, and then convert the dollars into Deutschmarks. Suppose the company chooses to issue a fixed rate bond, in dollars, and convert it to Deutschmarks in view of better rates or a higher appetite for new issues in the US market. Suppose the company is able to borrow at a coupon rate of 8%. The company can provide Deutschmarks to its subsidiary by changing the dollar into Deutschmarks at the prevailing exchange rate. However, if the exchange rate of Deutschmarks with respect to the dollar weakens, then the company will suffer. On every coupon payment date the subsidiary will have to pay more and more Deutschmarks to pay the interest to US investors in dollars. On maturity of the bond also the subsidiary will have to pay a higher amount in Deutschmarks. In order to avoid this currency exposure risk the company can enter into a cross-currency swap arrangement. For the sake of simplicity the company can take the dollar amount received from the issue of a bond and exchange it with a swap dealer who pays the firm the corresponding amount in Deutschmarks at the prevailing exchange rate. Now the company will have Deutschmarks and the swap dealer will have dollars. In the swap arrangement the swap dealer will agree to pay a coupon of 8% on the dollar amount. The company will agree to pay the swap dealer a predetermined fixed or floating coupon in Deutschmarks on the Deutschmarks amount. The dollars received from the swap dealer will, therefore, be sufficient to offset the payment of the dollar coupon. At the time of maturity the principal amounts will again be exchanged and the company will receive its dollar amount back to repay to its bond holders.

In this way the company has been able to convert a dollar-denominated loan into a Deutschmark-denominated loan in which it has its income flows. It has been able totally to avoid the currency exposure risk. The arrangement will work in the same way if the company decides to exchange the dollar amount received from the bond issue in the beginning with some other party. In that case the deal will be structured on a notional amount. However, in that case also the exchange of the principal amount at the maturity of the deal is essential in order to cover the exchange risk on the principal amount. An interest rate swap can be viewed as a special case of cross-currency swap in which both currencies are the same. Since both currencies are the same the net exchange of the principal amount required will be zero.

Options

All the instruments of financial engineering discussed so far (forwards, futures, swaps) provide immunity against future movements in market rates (Galitz, 1995). Whereas forwards and futures can provide this guarantee for months or a year, a swap can provide the guarantee for several years. Certainty, however, is not always the best thing. For example, a floating interest rate borrower will like to protect himself against a rise in interest rate but would welcome an interest rate fall. Similarly, a fixed rate borrower will feel happy if interest rates move up but would like to reap the benefits of a fall in the interest rate. None of the instruments discussed above provide this flexibility.

Options are unique among all the tools of financial engineering, for they give the buyer the ability to avoid just the bad outcomes, but retain the benefit of good ones. As such, options and all the products derived from them seem to provide the best of all worlds (Galitz, 1995). This flexibility, however, does not come free. A price has to be paid for it. A buyer or a seller in futures, forwards and swap transactions has an equal chance of gaining or losing, and therefore the expected value of the deal is zero. That is the reason why there is no up-front payment at the outset between the buyer and the seller.

The buyer of an FRA, or a future or a swap, simply enters into a binding agreement with the seller; they notionally shake hands on the deal but no entry fee need be involved. The market price is the fair price for both parties to the contract (Galitz, 1995). Options are different since they allow the buyer to benefit from the market movements in one direction, but not to lose in the case of market movements in the opposite direction. The seller of the option, conversely, can only lose and never gain. There is, therefore, no longer an expected value of a deal equal to zero. For this unequal position the buyer of the option has to pay a price to the seller of the option. In summary, what makes options different from all the other financial instruments is the asymmetry of the pay-off profile, and the consequent need for an up-front payment between buyer and seller.

Options have been in use, in one form or another, for several centuries. However, it was as recent as 1973 when the first organised market for share options was created, with the setting up of the Chicago Board Options Exchange. In 1978, the London Traded Options Market (LTOM) was opened on the floor of the International Stock Exchange in order to trade in options. Also in 1978, the European Options Exchange in Amsterdam was set up. The development of the options market was largely in response to the demand for risk-sharing and hedging instruments when there was a widespread unstable economic climate in the 1970s and early 1980s. The development of this market also contributed to the understanding of the factors relevant for options pricing.

Option contract

Dixon and Holmes (1992) define an option contract as that which:

> ... gives the holder of the contract the option to buy or sell shares at a specified price on or before a specific date in the future. The buyer of the contract pays the writer (or seller) for the right, but not the obligation, to purchase shares from, or sell shares to, the writer at the price fixed by the contract (the striking or exercise price).

Although this definition is specific to shares, in general terms one can say that in an option contract the writer of the option gives the right, but not an obligation, to the buyer to purchase from or sell to the writer something at a specified price within a specified period.

Parties to an option

There are two parties to an option: the buyer and the seller of the option.

- The *option seller* is the party who is obliged to fulfil the terms of the contract should it be exercised. This may involve delivering the underlying assets to the buyer for each option written. He is also known as the *option writer.*
- The *option buyer* is the party with the right to exercise the option. He has to pay for buying the option.

Call option

A call option is the right to buy a given quantity of an asset at a given price on or before a given date. This means that the writer (seller) of the call option must sell the underlying security to the buyer of the contract at the agreed price if the buyer decides to exercise the option before the given date.

Naturally, a rational buyer will exercise the option only if the market price of the underlying security is above the price agreed to in the option contract. Holding of such an option benefits only if there is a rise in the market price of the underlying asset above the contract price. The option contract just lapses if it is not exercised before the maturity.

Put option

A put option is the mirror image of the call option. A put option is the right to sell a given quantity of an asset at a given price on or before a given date. This means that the writer (seller) of the put option must buy the underlying security from the buyer of the contract at the agreed price if the buyer decides to exercise the option before the given date.

A rational writer of the option will exercise the option only if the market price of the security falls below the contract price. Holding of such an option will benefit only if the market price of the underlying security falls below the contract price. The option contract just lapses if it is not exercised before the maturity. This is why an option is considered to be a *wasting asset*. Since the option only has value for a fixed period of time, its value decreases, or 'wastes' away, with the passage of time.

Option strategy

Before discussing option strategies, it is useful to understand how an option contract works. Suppose there are two parties X and Y. X is the writer of the option and Y is the buyer. They enter into an option contract on the following terms:

- Y buys the call option for 3 (option premium) from X
- the underlying asset is one unit of ABC
- the exercise price is 100
- the maturity of the option is after 3 months from now
- it is a US-style option with a right for Y to exercise it at any time up to its expiration.

The term *premium* refers to the amount that the buyer of the option pays to the seller to acquire the option. The *exercise price* or *strike price* is the price at which the option can be exercised; this is normally fixed at the outset. The *maturity* or *expiry date* is the last date on which the option can be exercised.

At any time up to the expiration of the option contract, Y can decide to buy the asset from X and pay 100. Y will, however, do this only if it is beneficial to him. We will see below what are the conditions in which it will be beneficial for Y to exercise the option. However, whether Y decides to exercise the option or not, the option premium of 3 paid by him to X for the purchase of the call option will be retained by X.

If Y has bought a put option on the above terms, then he will have the right to sell the assets to X at the exercise price of 100 up to the expiration of the contract. Whether he exercises the option or not, the option premium paid by him (3) will be retained by X. Therefore the maximum amount that an option buyer can lose is the option premium. The maximum profit that the option writer can realise is the option premium.

The maximum cost (loss) to the option buyer if he enters an option contract is known but the gain to him in return is not. Similarly the maximum gain to the option writer is known but not the cost (loss). Under different option strategies, the gain potential to the option buyer is unlimited, but the loss potential to the option writer is also unlimited.

By combining the call and put options and the buyer and the seller of the contract, there can be four basic option strategies: buying a call option, writing a call option, buying a put option and writing a put option. Each of these four strategies is discussed below.

Buying a call option

Buying a call option is also referred to as *long a call option*. In the example above, Y has bought a call option at a premium of 3 at the exercise price of 100 on an asset ABC. Suppose the current market price of the asset ABC is also 100. To make it simple, let us assume that Y holds the option until its expiration date. We can now see how the profit or loss of Y changes if the price of asset ABC changes in the market. There are in fact five different outcomes:

- If the price of the asset is less than 100, then Y will not exercise the option. This is because he can purchase the same asset in the market for a lower price than the exercise price, which he will have to pay to X. Y will therefore prefer to lose the entire option premium paid. But this is the maximum amount that he can lose however low the price of the underlying asset ABC in the market.
- If the price of the asset ABC is 100, then there is no economic sense in exercising the option. In this case also Y will allow the option to lapse and lose the premium of 3.
- If the price of the asset ABC is more than 100 but less than 103, then Y will exercise the option and purchase the asset. Suppose the market price of asset ABC is 102, then Y can get the asset ABC from X by paying the option exercise price of 100. He can then sell it in the market for 102 and make a profit of 2. Since he has already paid a premium for the purchase of the option equal to 3, his net loss will be 1.
- If the price of the asset is equal to 103, then Y will exercise the option and realise the profit of 3, which will completely offset the cost of the option which he has already paid. This will be a no loss no gain position.
- If the price of the asset ABC is more than 103 then only Y can realise the net profit. Suppose the market price of ABC is 105, then exercising the option will generate a profit of 100 and a net profit of 3. As the price of asset ABC moves above 103 the gain to Y increases. If the market price of ABC is 113 the gain is 10, on a price of 123 the gain is 20, and so on. We can therefore say that for a price of 3 the gain potential to Y is unlimited.

Writing a put option

Writing a put option is also referred to as *selling a call option* or being *short a call option*. The profit and loss of writing a call option is exactly opposite to the profit and loss of buying a call option. In our example, therefore, the profit to X, the writer of the call option, on the expiration of the option is same as the loss to Y. Therefore, the maximum profit that X can receive is the option premium. At the same time, since the maximum profit to Y is unlimited, the maximum loss to X is also unlimited. The option will be equal to the highest price reached by the asset on the expiration date minus the option premium.

Buying a put option

Buying a put option is also referred to as being *long a put option.* In the example above, suppose Y has bought a put option at a premium of 3 at the exercise price of 100 on an asset ABC. Let us also suppose that the current market price of the asset ABC is 100. To make it simple, let us assume that Y holds the option till its expiration date. We can now see how the profit or loss of Y changes if the price of asset ABC changes in the market. The outcome can be one of five different types:

- If the price of the asset is greater than 100, then Y will not exercise the option because exercising the option would mean receiving a lower price for the asset, which can be sold at a higher price in the market. Y therefore loses the entire option premium paid. However, this is again the maximum amount that he can lose, regardless of how high the price of the underlying asset ABC goes in the market.
- If the price of the asset ABC is equal to 100, there is no economic sense in exercising the option. In this case also the option will lapse and Y will lose the premium of 3.
- If the price of the asset ABC is less than 100 but greater than 97, then Y will exercise the option and sell the asset to X. Suppose the market price of ABC is 98, then Y will gain by exchanging the asset ABC with X at the option exercise price of 100, as compared to the market price of 98. He will make a profit of 2 in the transaction. Since he has already paid a premium for the purchase of the option equal to 3, his net loss will be 1.
- If the price of the asset is 97, then Y will exercise the option and realise the profit of 3, which will completely offset the cost of the option, which he has already paid. This will be a no loss no gain position.
- If the price of the asset ABC is less than 97, then only Y can realise the net profit. Suppose the market price of ABC is 95, then

exercising the option will generate a profit of 95 and a net profit of 3. As the price of asset ABC moves below 97, the gain to Y increases. If the market price of ABC is 87 the net gain is 10, on a price of 77 the net gain is 20, and so on. Therefore, for a price of 3 the gain potential to Y is unlimited.

Writing a put option

Writing a put option is also referred to as *selling a put option* or being *short a put option*. The profit and loss profile of a short put option is the mirror image of the long put option. The maximum profit from this position is the option price. Theoretically the maximum loss will be generated when the price of the underlying security falls to zero. In this case the loss will be the strike price less the option price.

Summary

Buying calls or selling puts allows the investor to gain if the price of the underlying asset rises. Selling calls and buying puts allows the investor to gain if the price of the underlying asset falls.

Caps, floors, collars, swaptions and compound options

These are an important group of option instruments primarily to hedge the interest rate risk. Caps, floors and collars are among those instruments extensively used in project financing. An interest rate cap provides protection against an increase in interest rate, but at the same time allows the benefits of an interest rate fall to be enjoyed. For example, a borrower has taken a 5-year loan at LIBOR + 50 BP and has also bought an 8% 5-year interest rate cap. At each interest rate reset date, if LIBOR + 50 BP comes below 8% the borrower simply pays the prevailing market rate and takes advantage of the lower rates. If the interest rates on any reset date are higher than the cap rate, the cap will provide a pay-off to offset the consequences of the higher rate, effectively limiting the borrowing rate to the cap level.

An interest rate floor is used to limit the benefits from a fall in the interest rate once the floor level is reached. In practice, many users of interest rate caps seek to lower the cost of protection by selling a floor at a lower strike price. If the interest rate falls through the floor level on any reset date, the floor is exercised against the seller, who must pay the difference between the prevailing rate and the floor rate.

A collar is a combination of selling a floor at a lower strike rate, and buying a cap at a higher strike rate. It provides protection against a rise in rates and some benefits from a falling rate. A collar can be tailored to meet a compromise between interest rate protection and cost. By adjusting with the cap and floor rates, it is possible to create a zero-cost collar, for which no premium is to be paid.

A swaption is an option to enter into an interest rate swap on some future date. A payer's swaption is the right to pay the fixed rate on the swap, while a receiver's swaption is the right to receive the fixed rate.

Compound options are options on options. Just as we have an option on a swap (swaption), we can have an option on a cap (caption), floor (floortion) or collar (collartion). Compound options come in four possible categories: a call on call, a call on a put, a put on a call, and a put on a put. The first two give the holder the right to buy the underlying option and the other two give the right to sell the option. The underlying assets can be a call or a put. Galitz (1995) suggests that compound options should be bought primarily for two reasons. Firstly, to provide protection in a contingency situation when protection may or may not be needed; and, secondly, as a form of risk insurance that is cheaper than buying the option itself.

Options on futures contracts

Put and call options are being traded on an increasing number of futures contracts. Trading options on futures allows the speculator to participate in the futures market and know in advance what the maximum loss on his position will be. The purchase of a call gives the option buyer the right, but not an obligation, to purchase a futures contract at a specified price at any time during the life of the option. The underlying futures contract and the price are specified. The purchase of a put option gives the option buyer the right, but not an obligation, to sell a specified futures contract at a specified price. The profit realised with an option strategy is reduced by the option premium. The price of the option is determined in the same fashion that the price of an equity option is determined.

The tremendous versatility offered by options have led to a profusion of *second-generation options* or *exotic options*, which vary one or more of the conditions or features of standard options. Many of these products service a genuine need, and provide a valuable extension to the range of financial tools available for managing risk. Others are novel and innovative, but perhaps will prove to be short-lived. One or two are certainly interesting, but are probably solutions for which a problem has yet to be found (Galitz, 1995).

Asset-backed securities

The financial innovation of asset-backed securities (ABSs), or *asset securitisation*, during the 1980s dramatically changed the way of financing the acquisition of assets. In the traditional financing system a bank or an insurance company provides the loan and retains it on its portfolio, thereby accepting the credit risk, and seeks additional funds from the public to finance its assets. In ABSs a group of lending is packaged together and then issued as a new security the purchaser of which has a claim against the cash flows generated by the original lending. In ABSs more than one bank or financial institution may be involved in the lending capital. An insurance company may guarantee returns to investors in ABSs.

Asset securitisation has several benefits. It helps in obtaining a lower cost of funding. The capital is used more efficiently. It also diversifies the funding sources. The original lenders of the loan do not retain the credit risk. Their locked assets are released and thus can be used for more funding. This method is widely used for mortgage loans, home equity loans, loans by manufacturing companies and lease receivables.

Bibliography

Dixon, R. and Holmes, P. *Financial Markets: The Guide for Business.* Chapman and Hall, London, 1992.

Galitz, L. *Financial Engineering: Tools and Techniques to Manage Financial Risks.* Pitman, London, 1995.

Merna, T. and Dubey, R. *Financial Engineering in the Procurement of Projects.* Asia Law and Practice, Hong Kong, 1998.

CHAPTER FIVE

Restructuring projects

Introduction

Project finance provides a structure for financing large infrastructure projects, such as roads, bridges, power generation, telecommunication, water supply and other projects crucial to the social and economic development of a country. Project financing relies on a project's own revenues to repay lenders. It is based on a careful appraisal of the project's risks and returns and on sharing those risks and returns among parties involved in the project. In addition, a robust financial package has to be tailored around the project to ensure that repayment of debt is not compromised.

Recent experience, such as the 1997 Asian financial crisis, shows that projects procured under the project finance strategy are vulnerable to risks such as currency risk, as most of the projects are funded in foreign currency. To ensure these projects do not run the risk of default, they have to be restructured to get back on track.

Definition of restructuring

Restructuring is to pursue financially driven value creation using various financing instruments and arrangements (Pike and Neale, 1999). Restructuring may be regarded as the use of financial engineering techniques on an ongoing project. This means that new

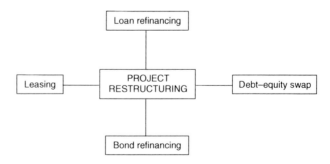

Figure 5.1. Restructuring techniques in project finance

ways of financing can be explored to enhance the value of the project, and to provide a reasonable rate of return to the investors.

If a project is having difficulties in generating sufficient revenues to service its debt and to maintain a reasonable rate of return, the promoter has to restructure its financing techniques to maintain its financial viability. In this chapter we discuss ways of restructuring in project finance, which include refinancing, leasing and debt–equity swaps. The restructuring technique in project finance is depicted diagrammatically in Figure 5.1.

Loan refinancing

Refinancing involves paying off an existing loan with the proceeds from a new loan, using the same property as collateral. There are two situations where promoters may consider refinancing. If the current interest rate (e.g. the London Interbank Offer Rate (LIBOR)) is lower than the rate on the debt, the promoter may consider refinancing so that short-term loans can be rolled over into longer term maturity loans. In infrastructure projects, most lenders are unwilling to offer long maturity in their lending. Thus, the promoter may take advantage of the lower interest rates to refinance short-term loans.

Lower than expected revenue generation may lead to projects already procured utilising project finance, thus defaulting on their debt obligations. Some projects have to be refinanced to ensure that debt obligations are met. However, this involves the risk of refinancing as the debt increases.

The International Finance Corporation (IFC) is the arm of the World Bank group that is actively involved in project finance by providing loans to the private sector. However, the institution recognises the need for long-term debt, which is difficult to arrange under project finance. So the bank provides loans to refinance under the following structure:

- the bank provides loans to the host government
- the promoter borrows short-term loans from commercial banks
- the host government undertakes to refinance the promoter's short-term loans through the provision of new loans at a later date.

In this way, instead of the promoter borrowing again from commercial banks to refinance, it takes out a loan from the host country. Since the host country is able to borrow at a lower interest rate, the promoter is able to refinance its short-term loans at a lower cost.

Refinancing can provide major financial benefits to a promoter organisation. Once the project has been constructed and operation begun, the lender no longer needs to worry about construction risks. By refinancing the promoter can achieve lower interest payments and thus increase his profit margin. Refinancing techniques have been used on a number of public–private partnership (PPP) projects in the UK resulting in greater profits to promoters. Public sector partners are opposed to additional profits being made by this technique. The authors suggest that if refinancing is specifically identified under the terms of the concession then the promoter should not be penalised for utilising creative financial engineering techniques to increase the commercial viability, in terms of profit, of the project. Under a partnership, pre-arranged agreements regarding the redistribution of any refinancing profit could be addressed to still maintain private sector incentive.

Bonds refinancing

In some cases, corporate bonds with a long maturity of up to 30 years and identifiable coupon payments can be issued to refinance short-term loans. Promoters can then compute their financial standings based on the fixed semi-annual or annual coupon payments.

For example, France Telecom is due to launch a close to US $16 billion bond issue to refinance its existing short-term and medium-term debt. The bulk of the multi-tranche bond will be issued in dollars, with a small part denominated in Euros and in Sterling, with maturity ranging from 2 to 30 years. To attract investors, France Telecom will increase the coupon on the bonds by 25 basis points for every notch that Moody's Investor Service or Standard & Poors lowers their rating below category A.

Leasing

Another way of restructuring is through leasing. The International Accounting Standards Committee defines leasing as:

A commercial arrangement whereby an equipment owner conveys the right to use the equipment in return for payment by the equipment user for a specified rental over a pre-agreed period of time.

The Equipment Leasing Association (ELA) in London defines a lease as:

A contract between a lessor and lessee for the hire of a specified asset selected from a manufacturer or vendor of such assets by the lessee.

The lessor retains ownership of the asset. The lessee has possession and use of the asset on payment of specified rentals over a period.

In project finance, there are principally two types of leasing: operating leasing and financial leasing. Under an *operating lease* the lessee (user) only leases the equipment to perform a specific job and the use of the equipment is only for a short period of time. The lessor (owner) undertakes to maintain and service the equipment under the terms of the lease. Since the lease is job-specific, the lease can be cancelled easily.

> An operating lease is any other type of lease – that is to say, where the asset is not wholly amortised during the non-cancellable period and where the lessor does not rely for his profit on the rentals in the non-cancellable period. (ELA, London)

In the case of a *financial lease*, the lessor does not normally provide for maintenance or service, although the lessor may provide maintenance for a fee. Unlike an operating lease, a financial lease cannot be cancelled due to its severe financial penalties that make cancellation not commercially viable. Also, the lessee has the right to renew the lease upon expiration. A financial lease is defined as:

> … a contract involving payment over an obligatory period of specified sums sufficient in total to amortise the capital outlay of the lessor and give some profit. (ELA, London)

Sale and lease back agreement

In a sale and lease back agreement the owners of the equipment sell the equipment and lease it back from the leasing company. This helps the owner to generate cash from the sale and at the same time use the equipment and pay periodic payments. An example of a lease and lease back agreement is a power project where a promoter can generate cash by selling some of its assets and leasing them back at the same time.

Leasing is being used increasingly in project finance, especially in financing of power projects. Fowkes (2000) noted that power plants are suitable for lease financing because of their long useful lives and their ability to maintain value, due to well-established operating characteristics and maintenance requirements.

In power projects, leasing removes the need for a substantial cash layout to acquire assets or equipment (peaking turbines, railcars, barges, etc.). The predictable lease periodic payments over the terms of the lease contract gives the promoter the ability to predict confidently the cash flows of the project.

Promoters who have strong incentives to maximise reported earnings usually adopt leasing as part of a project financing package because accounting expenses associated with ownership of the equipment are often larger than the accounting expenses associated with leasing. Accounting expenses associated with leasing are lower since promoters do not deduct depreciation of the equipment as an expense, as they do not own the equipment.

In developing countries, raising long-term finance at reasonable rates to purchase equipment proves to be a daunting task for promoters. Lease finance is an important financial instrument that can be used to reduce the cost of borrowing. In the UK the capital allowance of bodies such as passenger transport authorities can be used in leasing arrangements. The basic leasing mechanism might operate where a company has underutilised capital allowances. This could either be due to previous tax losses carried forward or very low nominal rates of tax. Through the leasing mechanism the company in effect sells this benefit to a financial institution which would normally pay tax at a much higher rate. The realisation of these tax benefits can in turn be reflected in a reduction in the cost of borrowing for the promoter. Leasing also provides the only form of medium- to long-term finance available for purchasing equipment.

In many countries, promoters can offset their full lease payments against income before tax, compared with just the interest on bank loans. Also, tax benefits associated with the depreciation of the equipment can be passed on to the promoters in the form of lower monthly payments.

An example of the use of leasing in project finance is the refurbishment of the HM Treasury building in the UK, which was designed and built between 1898 and 1917. In 1995, the then Conservative government decided to refurbish the building using the private finance initiative (PFI) strategy because the PFI scheme 'offers good value for money for the taxpayer, and provides modern, efficient office accommodation for the Treasury.'

Under the PFI scheme, a consortium known as Exchequer Partnership plc was formed as a special purpose vehicle (SPV) to bid for the work. The consortium comprised Bovis Ltd, Stanhope plc, Chesterton International plc, Chelsfield plc and Hambros Bank Ltd. This consortium was selected as the preferred bidder to refurbish the building and then lease it out to the treasury and other government departments for a certain period of time.

Negotiations between the consortium and the Treasury over the refurbishment of the building were terminated in 1997 due to the substantial expenditures and significant financial risks involved for the other government occupants of the building in terms of the disposal of property elsewhere.

However, in 1998 negotiations between the two sides were resumed to explore further whether a basis could be found under a PFI for developing proposals for the refurbishment of the Treasury building. In May 2000, a contract was signed with Exchequer Partnership plc to refurbish half the building for use by the Treasury. The building work started in July 2000, with completion scheduled for August 2002. Once completed, all Treasury officials will be housed in the new accommodation. The remaining space will then be refurbished and let by Exchequer Partnership plc to other public tenants and/or the private sector. Initial estimates of £158 million in refurbishment costs and a revenue generation of over £300 million over an operation period of 30 years was forecast for this project. Revenues would be generated from the private sector through rents and through rent from the Treasury.

Derived from this project was the segregation of the financial tendering from the technical tendering. This is now being suggested as the way forward for PFI projects.

Debt–equity swap

A debt–equity swap is another way of finance restructuring if a project faces difficulties in repaying debt. Projects that are facing financial difficulties but have great potential for growth usually adopt this technique. In a debt–equity swap debt is converted into equity. Converting debt into equity not only reduces the debt/equity ratio of a project, but also reduces the level of debt payments. Unlike debt, the promoter has no obligation to pay dividends when the project is not making a profit. A debt–equity swap can be used to convert debt or unpaid interest accruing from debt into equity.

For example, Thai Petroleum Industries, which was $3.7 billion in debt, restructured by swapping $756 million in unpaid interest for a 75% equity stake in the company. This effectively reduced the debt/equity ratio to 25/75.

The way in which a debt–equity swap works is shown in Figure 5.2. The bank assumes the debt incurred by the promoter and calculates the number of shares that can be converted from the debt based on the prevailing market price. The new shares will be issued at a value close to the value of the debt. The net result is a reduction in debt payment and a decrease in the debt/equity ratio. By swapping debt into equity, the promoter is in a better position to manage its debt and improve cash flow.

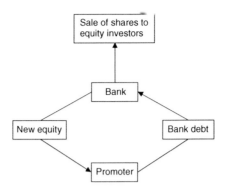

Figure 5.2. Illustration of a debt–equity swap

Besides debt or loans, bonds can also be swapped into equity to improve the balance sheet of the project. Through an investment bank, the bonds are bought on the open market at the prevailing market rate and converted into shares that can be newly issued in the stock exchange with the same market value. With less bonds available on the open market, the coupon payment is reduced.

Summary

In project finance every project is unique in the way it is financed, as lenders and investors view each project on an individual basis and analyse their expected returns commensurate with the risks involved. It is the promoter's task to plan and decide the type of financing instruments that are suitable for financing a project and, based on the financial analysis of the project, to present a case to the lenders and investors that such financing techniques are suitable for a particular project. In addition, it is suggested that restructuring techniques should be included in a financial package in case there is a need for such action, so lenders and investors know that their investments are protected.

The 1997 Asian financial crisis brought many privately financed projects to a standstill, and many projects had to be restructured. In this chapter we have discussed three ways of restructuring project finance, these being refinancing, leasing and debt–equity swap. These restructuring techniques are not exhaustive, but are considered to be most suitable and applicable in project financing.

In refinancing, the promoter can take advantage of lower interest rates to refinance its existing loans and to roll over short-term loans into longer maturity terms. On the other hand, a promoter may refinance its loans so as not to run into default. This increases the risk of refinancing as debt increases. Another way of refinancing is to issue corporate bonds, which have long maturity with fixed coupon payments.

The other method of restructuring is through leasing, whereby the promoter does not incur a lump sum payment for equipment purchase. Instead, periodic payments are made for the use of the equipment. The benefits associated with leasing make it well suited to project finance. Leasing can provide improved earnings and increase shareholders' values.

A debt–equity swap helps solve the liquidity problem of a project by converting some of the debt into equity. This reduces the debt/equity ratio of the project and also reduces the amount of debt repayment.

Bibliography

Dixon, R. and Holmes, P. *Financial Markets: The Guide for Business.* Chapman and Hall, London, 1992.

Fowkes, D. Leasing in project finance. *Journal of Project Finance,* **6**(1) (2000) 21.

Galitz L. *Financial Engineering: Tools and Techniques to Manage Financial Risks.* Pitman, London, 1995.

Merna, T. and Dubey, R. *Financial Engineering in the Procurement of Projects.* Asia Law and Practice, Hong Kong, 1998.

Pike, R. *Corporate Finance and Investment: Decisions and Strategies.* Prentice Hall, London, 1999.

Pike, R. and Neale, B. *Corporate Finance and Investment: Decisions and Strategies,* 3rd edition. Prentice Hall, London, 1999.

CHAPTER SIX

Financial markets

Introduction

A *market* can be defined as a place where buyers and sellers exchange goods and services in return for other goods and services. A financial market is a market where financial assets are exchanged or traded. French and Sward (1984) define an asset as:

> ... something that: (a) is owned by a specific person or firm, (b) confers or is expected to confer an economic benefit to its owners, and (c) can be given a value in terms of money.

Assets are further classified as either tangible or intangible. *Tangible assets* are described as physical objects over which someone has rights that confer valuable benefits to him, and an *intangible asset* as something that is owned by a specific person or firm, is useful, but has no physical form. The most important difference between a tangible and an intangible asset is their physical form. Buildings, land or machinery, the value of which depends on particular physical properties, possess tangible assets. The value of an intangible asset is represented through legal claims to their future benefits.

Financial assets are intangible assets. Their typical benefit or value is a claim to future cash. Financial assets are also referred to as *financial instruments* or *security*. The entity that has agreed to make the future cash payments is called the *issuer* of the financial instrument and the owner of the instrument is known as the *investor*.

Functions of financial markets

One of the fundamental roles of financial markets is to transfer funds between different economic units. The transfer of funds takes place between economic units, such as households, which have savings but no real investment opportunities, and economic units, such as firms or projects, which have productive opportunities but do not have sufficient funds to undertake these real investments. The role of the financial market is to bring these two groups together so that funds from the surplus sectors are channelled to deficient sectors. This is accomplished with the help of financial instruments and financial institutions. The interaction between buyers and sellers through financial instruments in the financial markets also performs the following functions:

- it helps determine the price of the traded instruments through the mechanism of demand and supply of the instruments
- it offers liquidity to the investor, by providing a ready market to sell an instrument he is owning, in return for cash
- it helps reduce the cost of transacting in financial instruments by providing a market where a counter party interested in purchasing the instrument could be easily located and by providing information about the merits of the instrument.

In an efficient market the price of a financial instrument reflects the aggregate information collected by all market participants.

Classification of financial markets

Financial markets can be classified in the following five ways:

- Classification by nature of claim:
  - *debt market*, a market dealing with debt instruments
  - *equity market*, a market dealing with equity instruments.
- Classification by maturity of claim:
  - *money market*, a market for short-term debt instruments
  - *capital market*, a market for longer maturity financial instruments.
- Classification by seasoning of claims:
  - *primary market*, a market for newly issued instruments
  - *secondary market*, a market for exchanging financial claims previously issued.
- Classification by immediate delivery or future delivery:
  - *cash or spot market*, a market in which a financial asset trades for immediate delivery
  - *derivative market*, a market for derivative instruments such as forward or futures contracts and options.
- Classification by organisational structure:
  - *auction market*, a market in which financial assets are traded by auction, such as in the New York stock exchange
  - *over-the-counter market*, a market in which a geographically disperse group of traders (linked via a telecommunications network) trade
  - *intermediated market*, a market in which trading through brokers takes place.

However, from the point of view of financing a project, financial markets may also be classified in the following broad categories: the money market, the bond market, the equities market, the foreign exchange market and the derivatives market. Each of these markets is discussed below.

The money market

The money market is the market for trading in short-term debt. Transactions are generally done electronically. In this market the maturity of instruments used varies from overnight to as long as 1 year. The size of deals can vary from the equivalent of US $250,000 up to US $50 million, and even more in many cases. Some of the typical instruments traded in the money market include the following:

- *Treasury bills*: governments often issue treasury bills for the purpose of raising funds. These bills constitute government credit and are virtually risk free. They are also known as *T-bills* or *treasuries*.
- *Certificates of deposits*: short-term loans to commercial banks, which can be of 3, 6, 9 or 12 months maturity.
- *Trade bills*: firms often issue trade bills to investors, with a promise to exchange them for a stated sum of money at a future date.
- *Banker's acceptance*: an agreement by a bank to pay a given sum of money at a future date.
- *Commercial papers/Euro-commercial papers*: short-term unsecured notes issued by corporations with a high credit rating and with a maturity of up to 270 days. Euro-commercial papers are issued in countries other than the country in whose currency the paper is denominated.
- *Euro-notes* (e.g. revolving underwriting facilities (RUFs) and note issuance facilities (NIFs)): short-term securities issued under note-issuing facilities.

For most of the above negotiable instruments, secondary markets exist that allow buyers and sellers to trade debt prior to maturity. For example, if an investor has purchased a 6-month certificate of deposit and subsequently he is in need of cash even before the instrument has matured, then he has the opportunity to sell the certificate to another investor in the secondary market and receive cash. The price at which these instruments are traded depends on a number of factors, including their time of maturity, their credit quality, the prevailing rate of interest and any accrued interest on the instrument.

This market also has a substantial amount of non-negotiable debt, such as inter-bank deposits, federal funds and local authority and finance house deposits, which are mainly used by banks and governments to meet liquidity requirements. They are, therefore, not discussed here.

The bond market

The bond market is segregated into the money market on the basis of maturity. Whilst most money market instruments have an original maturity of 1 year or less, bonds and notes, with maturity of more than 1 year, are issued in this market. In general, instruments issued in this market have a maturity in the range 2–10 years, but a maturity of up to 30 years is not uncommon. A number of bonds are perpetual in nature and have no fixed maturity date.

The biggest issuers of notes and bonds in most countries are central governments and local governments. Other than these, it is mostly large corporate entities that issue bonds. Instruments of the bond market may be classified as:

- *government bonds* – long-term treasuries issued by the government, having maturity above 1 year
- *corporate bonds* – fixed coupon rate bonds issued by corporate bodies
- *floating rate notes* – floating coupon rate bonds issued by corporate bodies
- *Euro bonds* – bonds issued in currencies other than the local currency of the country where the bond is issued
- *Euro-notes and medium-term notes* – notes issued in currencies other than the local currency.

The yield, and hence the price, at which bonds trade in the bond market depends on the level of interest rates prevailing in a particular currency. Yields usually differ for different maturites, giving rise to yield curves, which define the current yield for the respective maturity period. It is usually the case that yields increase for longer maturity, rewarding the investor for the additional risk involved in holding the bond with a longer maturity. One component of this risk is the chance that the bond issuer will default. Bonds issued by major governments are normally considered less risky, and set the base level for bond yields in a particular currency. On the other hand, bonds issued by other borrowers are considered to have a finite risk of default. Dollar- and Yen-denominated bonds are the largest, accounting for about two-thirds of the world bond market.

The equities market

Equity as a financial instrument is different from debt, there being no certainty about the stream of cash flows that the investor will receive. However, investors hold equity instruments in the hope of better returns in the future. It is also the case that equities normally outperform debt in the long run. The largest equity market, as regards turnover, is the New York Stock Exchange, followed by the London Stock Exchange and the Tokyo Stock Exchange.

Some of the instruments, such as convertibles, which share the characteristics of both a share and a bond, trade in the bond market until they are in bond form and, after conversion, in the equity market.

The foreign exchange market
This market has evolved out of the Bretton Woods agreement of 1944, which linked all the major currencies of the world with the US dollar through a system of fixed parities. This system worked well for almost two decades. However, due to significant differences in growth of different countries during the 1960s, it was no longer possible to protect the exchange parities. Eventually, in early 1970s the fixed parity system broke down and a floating exchange rate system was introduced. In this system the exchange rate is allowed to fluctuate continuously without any bound, depending on supply and demand conditions. This has necessitated management on a continuous basis of the foreign currency exposure of banks and corporate entities dealing in foreign currency. It has also provided the opportunity to speculators to speculate on the expectation of exchange rate movements (Merna and Dubey, 1998).

The foreign exchange market is different from both the debt and equity markets. Whereas a substantial proportion of transactions in the money, bond and equity markets is through stock exchanges, there is no such fixed location where transactions in foreign exchange take place and are recorded for settlement. The market is spread throughout the world in dealing rooms linked to each other through a web of telephones and computer networks. The three most important foreign exchange markets are London, New York and Tokyo, with turnovers of approximately US $300 billion, US $190 billion and US $130 billion, respectively. The volume of transactions worldwide is estimated to be around US $1100 billion. These are approximate estimates, because no one knows the exact volumes in the absence of any central clearing agency to record all the transactions.

The derivatives market
In all the markets discussed so far there is normally an exchange of cash from one party to another. The flow of cash is essential because

the borrower requires financing to fund various aspects of his operations. However, the transaction of cash exposes the two parties concerned, the borrower and the lender, to considerable risk. The development of derivative instruments and a market for them has provided a way to manage some of these risks efficiently. Derivatives, as the name suggests, do not exist in isolation. They are linked to some underlying instrument in the debt, cash or equity markets. For example, a currency option is linked to a particular set of currencies in the foreign exchange market, a bond future to a bond in the bond market, and a stock index future to a stock in the equities market. Derivatives are sensitive to fluctuations in the currency rates, interest rates or stock market, but these do not affect the underlying instrument. Since the risk on derivatives is smaller than that on the underlying asset, procuring a derivative product provides a relative hedge to the owner of that asset.

Globalisation of financial markets

Financial markets are no longer limited to national boundaries. Stiff global competition has forced national governments to deregulate or liberalise various aspects of the domestic financial markets. Now a project located in a particular country need not confine itself to raising funds from the local markets or depend on bilateral and multilateral aid. It can access funds from across the world at the cheapest rates possible to make the project competitive. Advances in telecommunications technology has made this possible by linking financial markets all over the world. This integration of financial markets throughout the world into an international financial market is known as *globalisation* of financial markets.

The financial markets available to a project in a country can be classified as either internal market or external market. The *internal market*, also known as the *national market*, is the market that operates within the boundaries of a country. It comprises the domestic market and foreign market. The *domestic market* is where issuers domiciled in the country raise their finances in the local currency. The *foreign market* is the internal market of a country in which issuers not domiciled in the country raise resources in the currency of that country. The raising of resources by a foreign entity not domiciled in that particular country is, however, regulated by the local laws and regulations. Nicknames have developed to describe various foreign markets in a country, such that the foreign market in the USA is called the 'Yankee market', that in Japan the 'Samurai market', that in the UK the 'Bulldog market', that in The Netherlands the 'Rembrandt market' and that in Spain the 'Matador market'.

The *external market*, also known as the *international market*, is characterised by the issue of securities, simultaneously to investors in a number of countries. The funds are raised in a variety of currencies, sorted out in the inter-bank market and made available to the issuer. The external market is also referred to as the *offshore market* or the *Euro-market*.

Market efficiency

As stated previously, the most important role played by the financial market is to transfer resources from resource-surplus sectors in the economy to resource-deficient sectors, even when there are two diametrically opposite interests operating in the market. The objective of the lenders is to maximise the returns on their lending with the least risk, and the objective of the borrowers is to minimise the cost of the capital raised subject to the condition that their full requirement of financing is met. That is why lenders prioritise their lending based on the return on the financial instrument available and the risks associated with it. On the other hand, borrowers prioritise their borrowing instruments based on the cost of raising funds through those instruments. The role of the financial market is not only to bring together the borrowers and the investors, but also to help share information about the risks and the returns of various financial instruments available in the market. For a market to be efficient, information about the securities and other factors that influence the security should be available easily, universally and at low cost. If this happens, then both investors and borrowers can be sure about the decisions they take regarding investing and borrowing, respectively.

Bibliography

Dixon, R. and Holmes, P. *Financial Markets: The Guide for Business*. Chapman and Hall, London, 1992.

Fowkes D. Leasing in project finance. *Journal of Project Finance*, **6**(1) (2000) 21.

French, D. and Sward, H. *A Dictionary of Management*, new revised edition. Pan Books, London, 1984.

Galitz L. *Financial Engineering: Tools and Techniques to Manage Financial Risks*. Pitman, London, 1995.

Merna, T. and Dubey, R. *Financial Engineering in the Procurement of Projects*. Asia Law and Practice, Hong Kong, 1998.

Pike, R. *Corporate Finance and Investment: Decisions and Strategies*. Prentice Hall, London, 1999.

CHAPTER SEVEN

The concession or build–own–operate–transfer (BOOT) procurement strategy

Introduction

The concession, or build–own–operate–transfer (BOOT), is a type of procurement strategy utilising project finance to fund infrastructure projects. This chapter defines the concept and structure of BOOT and explains the roles of the parties involved in a BOOT project. The chapter also discusses the conditions that are considered necessary for the successful implementation of BOOT projects. Three case studies are presented to illustrate the use of BOOT in the UK, China and Malaysia. The case studies also discuss the financial package used in the projects.

Private finance for infrastructure projects

A combination of factors, such as a rise in economic well-being and population growth, have led to growing demand for infrastructure services in both developed and developing countries. The inadequacy of public funds to keep up with the rising demand for infrastructure projects in both developed and developing countries has resulted in private finance being considered for an increasing number of infrastructure projects that would normally have been procured with public funds. In many cases low interest rate loans or deferred or subordinated loans are provided by government in conjunction with loans from the private sector for major infrastructure projects, resulting in a hybrid, mixed funding finance package. Concession contracts have been developed to facilitate private financing of infrastructure projects on the BOOT basis.

Early concept of BOOT

Although the term BOOT is relatively new, privatised infrastructure projects have been around for several centuries. The granting of a concession to the Perier brothers to provide drinking water to the city of Paris in the eighteenth century was the first initiated concession to be granted. The Trans-Siberian Railway and the Suez Canal, thought to be the first BOOT project in the modern world (Merna and Smith, 1996), were constructed, financed and owned by private companies under concession contracts.

The term *BOT* (*build–operate–transfer*) was first introduced in the early 1980s by Turkey's then Prime Minister Targut Ozal, and was known as *Ozal's formula*, although the concept was identified earlier when a privatised cross-harbour vehicular tunnel was first proposed in Hong Kong in the mid-1950s. There are other acronyms to describe concession contracts:

- FBOOT, finance–build–own–operate–transfer
- BOO, build–own–operate
- BOL, build–operate–lease
- DBOM, design–build–operate–maintain
- BOD, build–operate–deliver
- BOOST, build–own–operate–subsidies–transfer
- BRT, build–rent–transfer
- BTO, build–transfer–operate
- BOT, build–operate–transfer
- DBFM, design–build–finance–maintain
- ROT, rehabilitate–operate–transfer
- DBFO, design–build–finance–operate.

The most commonly used of these acronyms are BOT and BOOT, which are often used interchangeably. Many of the acronyms are alternative names for BOOT projects, but some projects differ in one way or another. There may be a distinction between BOOT and BOO with regard to the party in which the asset is vested during the concession period. For example, in a BOO project the promoter builds, operates and owns the facility. A BOO contract requires the promoter to finance, design, build and operate facilities over a certain period of time, but without the requirement to transfer the facility to the principal. In a BOOT project, the promoter is required to transfer the facilities back to the principal at the end of the concession period.

Although there are many acronyms, all concession contracts adopt the main functions of a concession strategy. In this book, the acronym BOOT will be used to cover all forms of concession contract.

There are a number of different contract strategies that can be utilised to procure a facility operated by the contractor or promoter for a specific period of time after commissioning. These strategies include turnkey contracts, BOO contracts and BOOT contracts.

Turnkey contracts

Under a turnkey contract strategy the operation and maintenance is often undertaken by the turnkey *contractor* for a period of time after commissioning to ensure that a fully operational facility is handed over to the *principal*. In BOO projects the promoter raises the finance and owns and operates the facility, usually for an infinite period. BOOT contracts require the promoter to finance, design, construct and operate facilities over a given period of time, the *concession period*, with the facility reverting to the principal at the end of that period. The difference between BOOT and BOO projects is the transfer element. In effect, under a BOOT strategy the principal remains the ultimate client or purchaser of the project.

The physical condition of the facility and the quality and quantity of the off-take at the time of transfer must be clearly identified in the *concession agreement* or in an appendix to it. This section considers the terms and provisions of the concession agreement relating to the transfer of the facility, and identifies the major risks and their effect on its overall commercial viability.

The BOOT strategy grew out of turnkey contracting – the supplier of complicated plant remained in charge of it as an operator, for a defined period of time, in order to train personnel and prove that the facilities could meet warranted performance specifications and capacities. Under the turnkey arrangement, operation elements evolved into longer terms and the supplier of plant and equipment realised a more substantial proportion of his reward from the operation phase. This period and the requirements are detailed in the *operation and maintenance schedule*. In some cases the operation and maintenance is extended or affermage (lease) contracts procured for the operation of the facility. In each case the principal approves the operation and maintenance requirements and their timing.

A *turnkey performance specification* indicates the required performance of a facility producing a defined off-take for a given period of time. Many turnkey contractors only require a performance specification, standards and conditions of contract from the principal's organisation to design, construct, commission and operate a facility that meets the principal's requirements. The turnkey contractor gives performance guarantees meeting the requirements of the principal, as defined in the performance specification.

In turnkey contracts, as in BOOT contracts, it is normal to provide only proven technology to meet the requirements of the contract specification. Experimental or untried technology is unlikely to be approved by lenders, since they are required to commit substantial sums on a limited recourse basis, where revenues are generated solely on the sale of off-take. Proven technology and historical operational data can be used to determine the operation and maintenance requirements of a new facility (Merna and Smith, 1996, vol. 2).

Concession contracts

The basis of any BOOT project is the concession contract. The concession contract is the primary contract for procurement of infrastructure projects using the BOOT strategy, and has been defined by Merna and Smith (1996) as:

> A project based on the granting of a concession by a Principal, usually a government, to a Promoter, sometimes known as the Concessionaire, who is responsible for the construction, financing, operation and maintenance of a facility over the period of the concession before finally transferring the facility, at no cost to the Principal, in a fully operational condition. During the concession period the Promoter owns and operates the facility and collects revenues in order to repay the financing and investment costs, maintain and operate the facility and make a margin of profit.

In a BOOT project, a project company, normally a special project vehicle (SPV), is given a concession to build and operate a facility that would otherwise be built by the public sector. The facility might be a power station, toll road, airport, bridge, tunnel, water supply and sewerage system, railway, communication or manufacturing plant.

The concession period is determined by the length of time needed for the facility's revenue to pay off the company's debt and provide a reasonable rate of return for its efforts and risks (UNIDO, 1996).

Structure of BOOT projects

Figure 7.1 shows a typical BOOT corporate structure (Merna and Smith, 1996).

The principal (host government)

The principal is the party who is responsible for granting a concession and eventually becomes the owner of the facility. The principal is usually the host government or a government agency. Most BOOT projects are structured without any financial or other forms of assistance from the host government. On some occasions, however, the government may furnish part of the land required for the project or provide some support loans to the project.

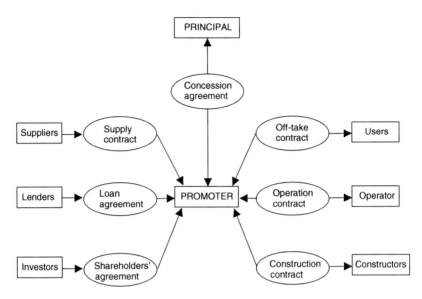

Figure 7.1. The BOOT corporate structure

For example, in the North–South Expressway in Malaysia, the government provided US $235 000 000 (about 13% of the total project cost) in loans towards construction costs, with a fixed interest of 8% per annum payable over 25 years (Tiong, 1990). In Australia, the government provided an interest-free loan of US $125 000 000 (about 23% of total project cost) to the project company, repayable over a period of 30 years for the construction of the Sydney Harbour Tunnel.

The use of the BOOT strategy for infrastructure development provides several advantages to the host government. The following are some of the reasons why host governments adopt the BOOT project procurement strategy (UNIDO, 1996):

- the use of private sector financing provides new sources of capital and reduces public and direct spending
- the development of projects that would otherwise have to wait, and compete for, scarce sovereign resources is accelerated
- the use of private sector capital, initiative and know-how reduces project construction costs, shortens schedules and improves operating efficiency
- project risk and burden that would otherwise have to be borne by the public sector is allocated to the private sector
- the involvement of private sector and experienced commercial lenders ensures an in-depth review as an additional sign of project feasibility

- there is technology transfer through the training of local personnel
- the development of national capital markets
- ownership of the facility reverts back to the host government at the end of the concession period
- there is an opportunity to establish a private benchmark against which the efficiency of similar public sector projects can be measured, and the associated opportunity to enhance public management of infrastructure facilities.

The promoter

The promoter is an organisation that is granted the concession to build, own, operate and transfer the facility back to the principal after the concession has expired. Winfield (in Merna and Smith, 1996, vol. 2) noted that promoters need to be financially strong because the investment and bonding obligation of concessions are far more onerous than those for conventional turnkey contracts. Also, the pre-contract expenditure can be quite considerable.

A selected promoter usually forms a separate entity (an SPV) to undertake the project. The promoter is responsible for borrowing funds to finance the project. The project company will enter into the necessary contractual arrangements with the suppliers, lenders, investors, users, operators and constructors.

Concession agreement

The concession agreement is the structured contract between the principal and the promoter. Merna and Smith (1996) define a concession agreement as:

> The contract between the principal and the promoter in which the concession is defined and granted and essential risks associated with it are addressed, described and allocated. The agreement will describe any facility vehicle, which the parties have agreed should be put in place to give effect to the concession, setting out its technical and financial requirements and specifying the parties relative obligations in relation to its design, construction, implementation, operation and maintenance over the lifetime of the concession.

The concession agreement is the centre of the web of contractual arrangements which, taken together, define the BOOT project. The agreement lists all the rights and obligations of the promoter and the principal, and the technical and financial requirements with regard to the construction, operation and maintenance of the project.

Off-take contract

There are two basic types of off-take contracts: the contract-led contract and the market-led contract. In a *contract-led project* a contract is entered into between the promoter and the user. The user may be an individual or an organisation who has agreed to purchase an agreed amount of output. An example of a contract-led project is a power station. The user enters into a contract with the promoter and agrees to purchase a certain quantity of electricity (kilowatt-hours) every month. An example of a contract-led project is the Shajiao Power Station in China, where the principal, the Chinese Electricity Authority, agreed to purchase 60% of the electricity output from the promoter.

In projects where the use of contract-led contracts is considered unfeasible, a market-led contract is used. In a *market-led project*, revenues are generated only when users are using the facility. An example of a market-led project is a toll road. Revenues are generated only when motorists pay to use the toll road. An example is the Malaysian North–South Expressway.

Supply contract

A supply contract is a contract between the supplier and the promoter. The supplier supplies the raw materials needed by the project to produce the required output. For example, the supplier could enter into a coal supply contract with the promoter to supply coal to generate electricity in a power station. The suppliers are often state-owned agencies or private companies.

Operation contract

Sometimes, when the project facility is not operated by the promoter, operation of the project is carried out by a specialist firm experienced in the operation of that particular type of project. The promoter enters into an operation contract with operators to operate and maintain the facility. The operation contract will set out the minimum operation and maintenance standards. Failure to perform to such standards may incur a penalty.

Loan agreement

The loan agreement is between lenders and the promoter, and states the conditions and terms of the loan facility provided to the project. Due to the size and complexity of a project that involves huge sums of money, one lender may not provide financing of the entire project. There may be more than one lender. Lenders are often commercial banks, pension funds, export credit agencies, international financial

95

institutions or insurance companies, which provide the loans in the form of debt to finance a BOT project. If there is more than one lender, there is usually a lead member. For example, in the Channel Tunnel project that linked the UK with France, there were 210 lending organisations, with NatWest as the lead bank.

Construction contract

The construction contract is the agreement between the promoter and the construction companies that build the facility. It may be a single company or a consortium of construction companies, especially if the project is too big for one company to carry out all the construction work. In some BOOT projects, the constructor assumes the role of promoter. An example is Dartford Bridge in the UK, where the construction company, Trafalgar House Group, assumed the role of the promoter along with several banks.

Shareholders' agreement

The shareholders' agreement requires some kind of equity investment. Investors provide equity to finance the construction of the facility and receive dividends as a form of return on their investments. Shareholders fall into two categories: those with a direct interest in the project (e.g. constructors, operators, suppliers) and those who are solely involved as equity investors (e.g. public shareholders, institutional investors such as insurance funds).

Potential of BOOT as a solution to governments' infrastructure funding problems

The BOOT strategy is being hailed as the procurement strategy utilising project finance to solve governments' infrastructure funding problems. But is it really the wonder solution? According to the World Bank, increased private sector involvement in infrastructure projects 'offers the twin benefits of additional funds and more efficient provision'. This is the argument in favour of BOOT – allowing the private sector to invest directly in infrastructure while relieving the government's financial burden.

However, not everybody agrees with this argument. Pahlman (1996) argued that the BOOT model is based more on free market ideology than on empirical evidence or fact. He reinforced the statement by continuing that BOOT has no track record and, until now, no major BOOT project has successfully completed all the stages (build, own, operate, transfer) according to the original plans. His views are supported by the fact that some BOOT projects have been either abandoned or delayed in the wake of the 1997 Asian financial crisis.

Quiggin (1998) stated that BOOT schemes are usually bad arrangements because companies that are good at construction are not necessarily good at operation and maintenance. He suggested that contracts such as operation and maintenance services be opened to scrutiny and competition after a period of time.

Guislain and Kerf (1995) pointed out that the BOOT concession contract is a flexible mechanism that can be designed to overcome a broad range of obstacles to private participation in infrastructure. An example of its flexibility is the option of leaving formal ownership of the project facilities to the host government. This is particularly useful in countries where the law or constitution excludes private ownership of certain infrastructure assets. However, they agreed that it is crucial to design a BOOT scheme that strikes a balance between the interests of investors, consumers and the host government, as well as matches the conditions of the sector and country concerned.

Klein and Roger (1995) stated that, although governments are able to borrow at a lower cost to fund infrastructure projects, private management of the facilities is better and more efficient, because of the lack of the ability of the government to balance the project's finance and the general budget. They claimed that state firms that receive budget subsidies to operate the facility have problems maintaining quality operations when fiscal problems arise.

According to the World Bank (1994) a 3% interest rate advantage of government spending would have to be offset by a cost saving of more than 20% for private ownership to be advantageous. Therefore, it is important to ensure that private management will yield sufficient benefits to offset the lower cost of sovereign funds.

Conditions for successful implementation of BOOT projects

The appraisal and implementation of BOOT projects are often time-consuming and very expensive exercises. This is why a promoter must be in a position to assess quickly whether the project is feasible under a concession strategy. The main conditions for the successful implementation of a BOOT project are: country, project and client.

Country
Economic stability
The economic stability of a country will affect the decision of whether to carry out or abandon a project. Parameters that affect the economic stability of a country include interest rates, inflation

and the strength of the local currency. Financial appraisal of a project is based on the assumptions of these parameters to develop a financial model to forecast the future revenues. These revenues are the only source of finance to repay the debt. If the economy of a country is unstable, for example if there are sharp interest rate fluctuations and high inflation rates, the promoter would have difficulty in developing an accurate financial model to make a forecast of future revenue generation. This would make the project financially unviable and investors would have no interest in investing in such a project.

Political will to carry out the project

If the project has been appraised and considered viable, the host government must have the political will to ensure that the project is carried out to completion. All parties to the project must show a high level of commitment. Host government support – legislative, regulatory, administrative and sometimes financial – is essential. The host government can provide various types of support to the project, such as special legislation or exemptions in the areas of taxation, labour law, immigration, customs, currency convertibility, profit repatriation and foreign investment protection.

Stock and capital markets

The host country should have a fairly well developed stock and capital market, wherein the equity and loans necessary to finance the project can be raised easily, as can any additional funds needed for further financing. However, there is a distinction between the financial instruments available in a developed country and those available in a developing country.

For example, a developed country like the UK is one of the largest financial centres in the world and has very well-established stock and capital markets. Therefore, raising of equity for BOOT projects is relatively easy, either by means of floating the project company on the stock market or through raising funds from private investors. Developing countries do not generally have such well-established stock and capital markets. Despite this drawback, BOOT projects can still be successfully implemented in developing countries, where returns can potentially be higher than in developed countries.

Legislative or judicial process

The legislative or judicial process of a BOOT project is quite complicated. The context may be outside the jurisdiction of the host government. Therefore, the laws and regulations under which the project operates must be recognised and allowed by law. For example, if there is no legislation regarding private ownership of infrastructure, special legislation may be required to authorise the private ownership and operation of BOT projects such as power plants, toll roads and water plants. In order to attract foreign investment and non-recourse debt financing, the host government must have a legal and regulatory system that is conducive to foreign investment.

If the project is to be realised and quality services are to be obtained at the lowest cost, the host government should also create clear processes for awarding BOOT contracts, and develop a framework for assessing bids and allocating project risks.

Project

The promoter must show that the project will generate sufficient revenue to repay the loans and provide a reasonable return to the investors. In a BOOT project, the promoter must show that the source of revenue is clear and certain and capable of providing a return on equity commensurate with the risks borne by the investors. Also, the project must be large enough to secure the development capital and time required by the promoter to generate revenues.

Client

The government is often the client in infrastructure projects such as power plants. When the client is the anchor purchaser of the project's output, such as electricity, it must be able to honour its payment to the promoter. This is crucial because such payments constitute the only source of income for repayment of project loans and dividends to banks and investors, respectively. The credibility of the government to honour its payment will boost investors' and lenders' confidence in the project.

Advantages and disadvantages of BOOT projects

As with other kinds of project procurement strategies, BOOT projects have both advantages and disadvantages associated with them. Merna and Smith (1996) list the following advantages:

- *Additionality*: BOOT projects offer the possibility of realising a project that would otherwise not be built.
- *Credibility*: the willingness of equity investors and lenders to accept the risk indicates that the project is commercially viable.

- *Efficiencies*: the efficiency of the promoter and its economic interest in the design, construction and operation of the project will produce significant cost efficiencies to the principal when the concession period ends.
- *Benchmark*: the project can be used as a benchmark to measure the efficiency of a similar public sector project.
- *Technology transfer and training*: the continued involvement of the promoter throughout the concession period will provide the necessary technology transfer that will benefit local industry and people.
- *Privatisation*: a BOOT project will help in a government's policy of infrastructure privatisation.

The disadvantages of BOOT projects include:

- *Additionality*: commercial lenders and export credit guarantee agencies will be constrained by the same country risks.
- *Credibility*: there will be no credibility if the government provides too much support to the promoter.
- *Complication*: a BOOT strategy is a highly complicated structure that requires detailed planning, time and money throughout the concession period. The promoter must have the commitment and interest to maintain the project.
- *Viability*: few BOOT projects actually reach the construction stage due to the lack of experienced developers and equity investors, the ability of governments to provide support and the workability of corporate and financial structures.

Typical operation and maintenance issues in BOOT projects

In BOOT projects the operator assumes responsibility for maintaining the project assets and operating them on a basis that maximises the profit or minimises the cost on behalf of the promoter. The operator, like the constructor, may also be a shareholder in the promoter.

It is important to recognise that the success of a BOOT project is based on providing an effective service to the client. To achieve this, operation and maintenance requirements need to be addressed comprehensively at the bidding stage of a project and not as an afterthought once the project has been awarded. The operator should not only have the required operation and management experience to suit the project, but also the financial strength. The issues that need to be determined at tender stage include:

- the type of operation and maintenance contract that would suit the facility
- the operation of the technology over the full working life of the facility
- the costs and resources over the operation period
- the demand of the facility
- the method of operation and maintenance required by the principal
- licences, authorisations and permits
- performance, liabilities and guarantees
- secondary contracts
- the method of revenue collection and currencies
- the availability of operational materials
- the type and time-scale of major maintenance requirements
- operational risks.

Concession periods vary between 10 and 20 years for process and industrial facilities and 20 and 55 years for infrastructure projects. Many process and industrial facilities are of a dynamic nature and may require a major operation and maintenance commitment throughout the concession, and possibly even extending into further arrangements for the operation of the facility after transfer to the principal.

Some infrastructure facilities require minimal operational input, the primary concern being the collection of tolls from users, but do require major maintenance programmes at defined intervals over the project lifetime. Process facilities, such as water treatment plants, may comprise a reservoir, treatment facility and metered distribution network, with each individual component requiring specialised operation techniques. The resources required, in the form of power, chemicals, manpower and the collection of revenue, results in complex operation and maintenance programmes. The annual cost of operating static infrastructure facilities can often be as low as 0.25% of the total construction cost, whereas the annual cost of operating a process facility can be as high as 10% of the total construction cost.

The value of including an operator to assess and cost the performance and demand risk at the bidding stage is paramount to the success of a project. If the promoter has not worked in a turnkey environment and has little experience of operating and maintaining a facility, it may be advantageous to enter into a contract with a specialist operator for the particular type of facility. Many specialist operating companies, such as toll road operators, are subsidiaries of major construction organisations. Several of the promoter organisations currently bidding for motorway concession contracts have incorporated either Italian or French operators for this reason. Operators may be specialist organisations interested in operating the facility after commissioning, existing operational organisations, or organisations specifically formed, usually by the promoter, for individual projects.

Many promoter organisations require operators to provide equity in the concession company. This provides the promoter with a guarantee that the operator will perform over the operation period, and provides the operator with greater returns through dividends. In other cases, operators may be required to provide operating capital over the early operation period prior to revenue generation. Performance bonds may be required by the promoter or the principal as part of the operator contract; similarly, operator guarantees may be required by the lenders. In a number of developing countries opposition to granting licences to foreign operators has proved problematic as no local operators exist.

The choice and efficiency of the operator is vital to the success of the entire project. In a power plant, deficiencies in the ability of the operator may result in the plant falling into disrepair through neglect or negligence (e.g. boiler explosion). Operator inadequacy may also affect the downtime allowed by the principal and result in performance penalties and loss of revenue from the off-taker.

The operation of a facility may be affected by environmental constraints imposed by the host country government during the operation and maintenance phase. In the case of operators involved in projects within the European Union, where there are already more than 150 environmental measures under review, the ability to cope with possible changes in environmental legislation could be the key to the viability of the concession.

Operation and maintenance contracts within BOOT projects

The operation and maintenance phase of a BOOT project is crucial to the success of the project, because it is during this phase that revenue is generated. The form of operation and maintenance required by concessions varies from project to project. The type and location of the facility will determine the operation and maintenance strategies to be adopted. With these types of facility, the supply and off-take contracts, and hence revenue generation, are inextricably linked to the operation of the facility. In infrastructure projects the collection of revenue is usually on a direct basis with the user.

Operation and maintenance contracts undertaken by a specialist operator are usually on a fee basis, and the operator is unlikely to give full indemnity to the promoter in the event of negligence or failure by its employees or agents. In power generation projects the operator is often compensated on a cost-plus basis, with bonuses or liquidated damages payable depending on performance levels. Most operators, like contractors, prefer a service contract providing profit for minimum risk with no responsibility for revenue. Operators will typically prefer to have a contract with the promoter providing for payment of services rendered, usually on a cost-recovery and incentive basis.

Many operation and maintenance contracts are based on model forms of construction contract, because these forms of contract are well understood and legal precedents as well as case law exist to aid in their interpretation. Specially drafted contract documents may sometimes be preferred, as there could well be legal ramifications in using a contract out of its intended context. The contract must also go further than the operation and maintenance of the hardware (e.g. plant, equipment structures). It must also cover the software or organisational aspects of planning, stock control, billing and revenue collection, customer and employee relationships, training and career development. The training element of the operation and maintenance of a BOOT project becomes critical to the principal as the concession approaches transfer.

Operation, maintenance and training contracts may be in the form of operation only, maintenance only, training only, operation and maintenance, operation and training, maintenance and training and operation, maintenance and training:

- *Operation only*: under this form of contract the operator would undertake to operate and manage the facility, with maintenance carried out by the promoter's organisation. An example of this form of contract may be considered for a toll road, bridge or tunnel, where the operator would be primarily concerned with revenue collection but not competent to carry out maintenance works.
- *Maintenance only*: in this case the operator would be responsible for the maintenance of the facility only, with revenue collection being carried out by the promoter's organisation. Examples of this form of contract may be the maintenance of a road or particular equipment, which may be required due to wear and tear throughout the concession period.
- *Operation and maintenance only*: under this form of contract the operator would undertake to operate the facility and perform all routine and non-routine maintenance, and return the

facility to the principal in full working order at the end of the concession period. In addition, the operator would be responsible for the provision of spares and consumables during the operation period. An example of this form of contract is the operation and maintenance of a water treatment facility or process plant.

- *Operation, maintenance and training*: here the operator is obliged to train personnel to operate and maintain the facility until the end of the concession period, so that fully trained personnel are transferred with the facility to the principal, who may employ such personnel to train his own operators. This form of contract may be considered for process or industrial plants, which require high levels of operation and maintenance skills in order to ensure that revenue is generated during and after the concession period.

In a concession, the contractual arrangements are less clear with respect to the operation, maintenance and training of operatives for such projects. Promoters of BOOT projects considering procuring the services of specialists in operation, maintenance and training need to consider the following:

- *Time*: the length of the operation period. In the case of a BOOT project the concession period is stated in the concession agreement, but these time periods are usually longer than a promoter is willing to commit to a single operations contract.
- *Involvement*: the division of responsibility for the provision of all resources relating to the operation and maintenance between the operator and the promoter.
- *Specification*: the level of restraints required in the operation and maintenance specification, and the criteria against which performance will be assessed.
- *Payment*: payment mechanisms, such as cost plus, lump sum, unit rate, schedule of rates, performance, equity and percentage of profits.
- *Training*: the balance between the training of staff and the transfer of the facility to the principal.
- *Contract*: the type of contract, the existing model form of contract or a special draft.

This list is by no means exhaustive, and the final choice will depend on many variables, including the type of work to be carried out, available budgets and the political, social and economic characteristics of the organisations involved and the country in which the BOOT project is being constructed.

Risks in the operation and maintenance phase of BOOT projects

The risks associated with the operation and maintenance phase can be extensive and costly, particularly if the revenue generation of the project is adversely affected. As the project relies solely on the revenues generated to service debt, pay premiums to equity investors and create a reasonable margin of profit, any risks that reduce revenue receipt in the operation period are particularly significant. The early appraisal of operation and maintenance risks is particularly crucial if the promoter is inexperienced in the operation of the particular type of facility.

Additional risks can result from the operation of an existing facility undertaken as part of the concession. Since this usually provides early income, the operator must ensure that the existing facility remains fully operational and determine all maintenance requirements over its operating life.

Documented problems affecting the operation of BOOT projects have included:

- cash flow problems at start up
- technical failure causing delay
- lower production than projected
- poor management of the facility
- technical obsolescence
- increased prices of or shortages in raw materials and consumables
- foreign exchange, currency and inflation problems
- government interference
- *force majeure* during the operation period.

The costs associated with meeting the entire operation and maintenance requirements right up to transfer of the facilities must be clearly identified at the tender stage. The transfer element of a BOOT project should be addressed by both the principal and the promoter, and provisions should be specifically drawn up to cover all the requirements and obligations of each party prior to and during the transfer period and final hand-over of the facility. Any general, specific or common term or project condition identified in the concession agreement that may affect the transfer of the facility should be determined.

The identification and allocation of risks and the means by which the facility will be transferred and assessed should be determined and agreed through the provisions of the concession agreement, in order to alleviate the possibility of conflict between the parties to the concession and with secondary contracts associated with finance and operation.

Methods of financing, amortisation periods and possible additional funding over both the concession period and the project life cycle must be explored to determine the effect on the transfer of part, or all, of the facility. The principal should also compare the costs of adopting a BOO strategy (instead of BOOT), if it is considered that the facility will either be obsolete or require a high capital expenditure at the time of transfer.

BOOT case studies

In this section we discuss the structures and features of three BOOT projects that are in operation:

- Dartford Bridge, UK, $310 million
- Shajiao Power Plant, China, $517 million
- North–South Expressway, Malaysia, $1.8 billion.

Dartford Bridge, UK

The Dartford Bridge is the third River Thames crossing at Dartford, joining the M25 orbital motorway. The government recognised an urgent need for an additional crossing as the capacity of the existing

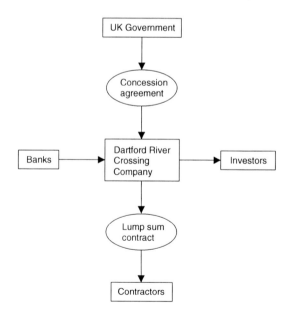

Figure 7.2. The BOOT structure for the Dartford Bridge project

Table 7.1. The finance structure of the Dartford Bridge project

Source of funds	Amount
Bank of America and other commercial banks	$185 million in term loan $18 million in stand-by facility $18 million in working capital facility
Prudential Assurance Company and other institutional funds	$64 million in subordinated loan stocks at 16 years $57 million in subordinated loan stock at 20 years
Dartford River Crossing Company	$1800 in nominal equity (pinpoint equity)

tunnels was exceeded by some 25 000 vehicles per day in 1987–1988. The promoter of this project is a consortium comprising several banks and Britain's Trafalgar House Group. Together, they form the Dartford River Crossing Company. The BOOT structure of the project is shown in Figure 7.2.

The concession period for the project is a maximum of 20 years, and the ownership of the project will be transferred to the British Government either at the end of the period or when all accumulated debt has been repaid, whichever is earlier. The concession period started in 1988 and will end in 2018. Under the concession agreement, the promoter had to purchase two existing toll tunnels at a cost of $80 million and to clear the existing $88 million (about £55 million) debt on the tunnels carried by the Kent and Essex County Councils. This was considered to be a good arrangement, as the two existing tunnels are making a profit. At the same time, the promoter can earn toll income from the tunnels and thus reduce the initial financing requirements, thus allowing immediate payments to be made to the institutional investors (Tiong, 1990).

Loans were arranged and provided by several banks led by the Bank of America. Institutional investors led by the Prudential Assurance Company raised most of the funds. A feature of this project is that there is no equity contribution. The promoter only provided a nominal equity of US $1800. This is known as *pinpoint equity*. Merna and Smith (1996) observed that the lack of equity illustrates the lender's confidence in the success of this project. The financing structure of the project is shown in Table 7.1.

The Dartford Bridge project is one of the most successful BOOT projects in Britain. The bridge was completed on time and within budget, due to the careful control of construction costs by the management team. In this project the main source of revenue was the toll collected from motorists. Accurate projections of traffic flows played a major part in the success of this project. In 1998–1999 a total of 50 420 231 vehicles passed through the tunnels and over the bridge, a daily average of 138 137 vehicles. Since the bridge was opened, there has been an increasing volume of traffic, thus ensuring a steady stream of revenue. Table 7.2 shows the total traffic flow through the tunnels and over the bridge since 1991.

To ensure that infrastructure projects (bridge, toll road) are commercially viable, the revenue stream must be clearly identified. As this is a market-led contract, the risks are that motorists may not want to pay for the use of the bridge. However, in the case of Dartford Bridge, the promoter realised the need for another river crossing at Dartford, as it is the vital link in the M25 motorway, which is considered Britain's most important orbital road. The initial projection for traffic flows also turned out to be correct, with the daily average numbers of vehicles using the bridge far exceeding the promoter's projection (see Table 7.2).

Shajiao Power Plant, Guangdong Province, Republic of China

Shajiao power plant is a 2×350 MW coal-fired power station in the Guangdong province of China, and is the first power station to be procured on a BOOT basis in China. Rapid economic developments in the Guangdong province had led to a vast expansion in infrastructure projects, such as roads, bridges, railways and ports. This, is turn, required an expansion of electricity generation facilities. There was great demand for electricity for further developments. At that time, China did not have the resources or expertise to finance the development of a power station, and it therefore had to rely on foreign investment. Hopewell Holdings of Hong Kong signed a concession

Table 7.2. *Traffic flows over Dartford Bridge and through the Dartford Tunnels*[*]

| Year | No. of vehicles | | |
	Total	Daily average	Highest daily throughput
1991–1992	34 797 684	80 440	128 047 (28 August)
1992–1993	37 385 483	95 076	135 351 (27 August)
1993–1994	39 947 382	102 426	144 728 (26 August)
1994–1995	42 557 309	109 445	148 088 (25 August)
1995–1996	44 363 898	116 596	153 156 (30 August)
1996–1997	46 403 105	121 213	161 734 (1 August)
1997–1998	48 455 901	127 132	169 098 (28 August)
1998–1999	50 420 231	132 756	174 368 (28 May)

[*]Source: Dartford River Crossing Company

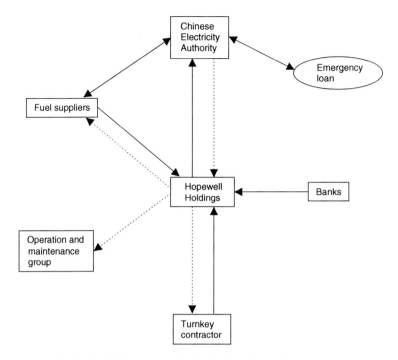

Figure 7.3. The BOOT structure for the Shajiao Power Plant project

agreement with the Chinese Electricity Authority to design, construct, test, commission and operate a power station. Figure 7.3
shows the BOOT structure of the project.

The concession period for this project was 10 years, but that does
not include construction time. The concession period was from
1987 to 1997. In this project, the Chinese Electricity Authority
agreed to purchase a minimum of 60% of the electricity from the
plant on a take-and-pay basis and also agreed to pay a fixed price per
kilowatt-hour over the concession period. This effectively determined the revenue stream for the project. The authority also agreed
to arrange for the supply of coal for the whole of the concession
period, at a fixed price per tonne.

The power plant was 100% financed in foreign currency. Hopewell negotiated for half of the electricity price to be paid in Chinese
Renminbi to pay for the Chinese coal. The promoter contributed
some equity to the project and the rest of the funds were borrowed
offshore from 46 international banks. The promoter also negotiated
deferred credits from the construction consortium, allowing for repayment over a 7.5-year period to ease the cash flow of the project.
The finance structure of the project is shown in Table 7.3.

Table 7.3. The finance structure of the Shajiao Power Plant project

Source of funds	Amount
International banks	$500 million in term loans
Hopewell Holdings	$17 million in equity

The plant was completed, fully tested, commissioned and in full operation within a period of 33 months, 6 months ahead of schedule. The successful completion of the project was due to good engineering design, efficient site supervision and a dedicated management team.

The concession for this project has ended and it has turned out to be a great success, both for the Chinese Government and for Hopewell Holdings. The promoter has been making profits since the start of the concession period through the sale of the electricity to the Chinese Electricity Authority. Walker and Smith (1995) cited the factors that contributed to the success of this BOOT project as:

- Gordon Wu, Chairman of Hopewell Holdings, who was able and willing to promote the scheme and to lobby effectively for its construction (i.e. the existence of a project champion)
- the willingness of government officials to co-operate
- the use of tried and tested turbine and construction technology.

The authors consider the most important factor that contributed to the success of this project was the take-and-pay contract between Hopewell and the Chinese Authority. The Chinese Authority agreed to purchase 60% of the electricity output, and this has provided a steady stream of revenue for the promoter. Furthermore, the promoter knew that Guangdong province had a shortage of electricity, and that the remaining 40% of the electricity output would soon be purchased. According to Walker and Smith (1995), the province would have lost some $500 million in economic value due to factory closures through power shortages if the power plant had not been built on time. This further cemented the need for the power plant to be built.

The power station has recently been handed back to the Chinese Electricity Authority in full operation, and there are plans to build more power stations in Guangdong province to meet the increasing demand for electricity.

North–South Expressway, Malaysia
The 30-year concession contract was awarded in 1988 to the United Engineers (Malaysia) Berhad, who then formed another project

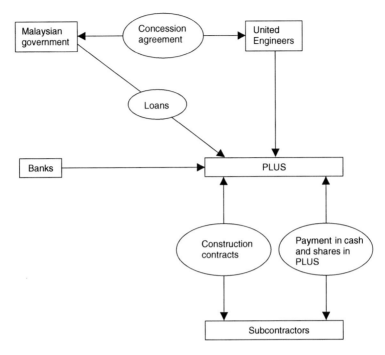

Figure 7.4. The BOOT structure for the North–South Expressway project

company called Project Lebuhraya Utara Selatan Berhad (PLUS) to design, construct, finance and operate the expressway. The toll road formed part of the 800 km North–South Expressway, which stretches from the Thai border to Singapore and runs through seven west-coast states, and links 40 major towns and cities in the Peninsular Malaysia. The award of the contract was in line with the Malaysian government's privatisation programme, which was embarked in the early 1980s and was aimed at involving the private sector in the management of public assets. Figure 7.4 shows the BOOT structure of the project.

Under the concession agreement, the promoter would take over the operation of the existing 377 km of tolled Federal Highways 1 and 2 to provide an early income. The project was financed through a conventional debt–equity structure. Most of the finance required for the project was raised offshore in Hong Kong, Singapore and London on a limited recourse to the Malaysian government. The government also provided a $235 million support loan to the promoter. The promoter paid its subcontractors 87% of the contract values in cash and the remaining 13% in equity shares in the project company, which is listed on the Kuala Lumpur Stock Exchange. In

Table 7.4. The finance structure of the North–South Expressway project

Source of funds	Amount
Offshore banks	$900 million in term loans
Malaysian government	$235 million in support loans payable over 25 years at a fixed interest rate of 8% per annum, with a 15-year grace period
PLUS	$9 million in equity
Shareholders	$180 million in equity

addition, the Malaysian government provided the promoter with the guarantee that it would make up the shortfall if the local currency fell more than 15% against the rates at the time of draw down of funds. The finance structure of the project is shown in Table 7.4.

The project was completed 15 months ahead of schedule and the expressway has halved the travel time for inter-urban journeys on the west coast. The expressway is aimed at providing major economic benefits to Malaysia by increasing industrial productivity, facilitating access to ports and airports and promoting domestic tourism.

However, the financial crisis in 1997 greatly affected the parent company, PLUS, Renong Berhad. This forced PLUS to restructure its debt with the help of the Corporate Debt Restructuring Committee (CDRC). In 1999, under the debt restructuring scheme, PLUS issued a 7-year zero coupon bond valued at about US $2 billion (RM 8 billion) to settle the claims of the creditors of Renong and United Engineers.

The financial crisis in 1997 did not help this project at all. The depreciation of the Malaysian Ringgit against the dollar has increased the promoter's debt burden and, although the economy is recovering, the promoter was left in a dire financial position. However, given the current economic situation in Malaysia and barring any crisis, together with substantial government support, the project should be successful in the long term.

Bibliography

Arndt, R. and Maguire, G. *The Risk Identification and Allocation Project: A New Frontier In Understanding Project Risk Allocation.* BenchMark Issue, pp. 5–7. Victoria Department of Treasury and Finance, Melbourne, 1999.

Guislain, P. and Michel, K. *Concessions – The Way to Privatize Infrastructure Sector Monopolies.* World Bank Public Policy for the Private Sector, Note No. 59. World Bank, Washington, DC, 1995.

Keong, C. H., Tiong, R. L. K. and Alum, J. Conditions for successful privately initiated infrastructure projects. *Proceedings of the Institution of Civil Engineers*, **120** (1997).

Klein, M. and Roger, N. *Back to the Future: The Potential in Infrastructure Privatisation*. World Bank Public Policy for the Private Sector, Note No. 30. World Bank, Washington, DC, 1995.

Merna, T. and Njiru, C. *Financing and Managing Infrastructure Projects*. Asia Law and Practice, Hong Kong, 1998.

Merna, T. and Smith, N. J. (eds). *Projects Procured By Privately Financed Concessions Contracts*, vols 1 and 2. Asia Law and Practice, Hong Kong, 1996.

Pahlman, C. Build–operate–transfer (BOT) private investment in public projects … or just more public subsidies for the private sector? *Watershed*, vol. 2(1). Ecological Recovery and Regional Alliance, TERRA, Bangkok, 1966.

Quiggin, J. Private infrastructure options: BOOTs, BOTs and BOOs. Presented at: *BOOT: In The Public Interest?*, University of Technology, Sydney, 1998.

Tiong, R. L. K. Comparative study of BOT projects. *Journal of Management in Engineering*, **6** (1990).

Tiong, R. L. K. Competitive advantage of equity in BOT tender. *Journal of Construction Engineering and Management*, **121**(3) (1995).

UNIDO. *The Guidelines for Infrastructure Development Through Build–Operate–Transfer (BOT) Projects*. United Nations Industrial Development Organisation, Vienna, 1996.

Walker, C. and Smith, A. J. (eds) *Privatized Infrastructure – the BOT Approach*. Thomas Telford, London, 1995.

World Bank. World Development Report. Washington, DC, 1994.

CHAPTER EIGHT

The private finance initiative

Introduction

The private finance initiative (PFI) was introduced in the UK in November 1992 by, the then Chancellor of the Exchequer, Norman Lamont. It came against the backdrop of a major programme of privatisation over the 1980s under the leadership of, the then Prime Minister, Margaret Thatcher. Over this period many state owned and operated organisations, such as British Telecom and British Gas, were sold off to the public sector, in many cases floated on the Stock Exchange. This was in the drive to free up the economy, allowing market forces to dictate and reduce public sector inefficiencies and overspending.

Conceptual standpoint

Traditionally, the selection of a procurement strategy was driven by the need to acquire an asset. The emphasis of the PFI is not the acquisition of an asset but the procurement of a service. The route taken to achieve this end is infinitely variable, but the concept is broadly unchanged – the private sector will provide funding for the capital project and operate the facility to provide a public service. Revenue will be achieved from this, either directly from the user, or through a payment mechanism negotiated with the public sector.

The only reason why governments occupy buildings is to help them deliver services or achieve policy objectives. What counts is the quality of the service provided from those buildings. Having a private provider of a capital asset can enable public bodies to purchase services without the need for the initial capital investment. Therefore the PFI seeks to obtain the service, and how this is achieved is dictated by the ingenuity of the private sector. The private sector will provide the funding for the capital projects and operate a facility for the public benefit. It will receive revenue from operating this service and hence make a profit.

The origin of the PFI

Historically the public sector has procured capital assets, on which they base their services. Government has only procured assets in order to supply a service from them, but current political thinking places greater emphasis on the service rather than the asset. The PFI was a policy born out of a series of privately financed projects, beginning with the Channel Tunnel project in 1987. Other transport orientated projects swiftly followed, such as the second Severn Crossing, the Dartford Bridge, the Skye Bridge, the Manchester Metrolink and the London City Airport. The PFI has developed as an alternative method of procuring services for the public sector.

Why the PFI?

The underlying principle behind the introduction of the private sector has many dimensions. The obvious one is in the pure PFI case, where a facility and service are provided at minimal cost to the public sector. An example of this is the second Severn Bridge, which can be held up as the perfect PFI project. A second dimension is the exploitation by the public sector of the private sector's ability to design and manage more efficiently. The public sector is characterised by a poor track record of integrating the design, build and operation of assets that it uses to deliver (or facilitate the delivery of) public services.

The role of the PFI has often been debated. Is it a substitute for or a complement to government spending? This question will be considered in a later section. Whatever its role, any reduction in capital expenditure needs to be made up by PFI schemes. For instance, cuts in UK government spending amounting to £2.6 billion in 1997 were compensated for by £2.1 billion in PFI provision. Warnings were made by the construction industry that the government must not become 'overdependent' on the PFI. The recent pan-European downward pressure on public capital budgets has forced governments to rethink their procurement strategies, with PFI-type schemes becoming more popular throughout Europe.

A privately financed public service is not a new concept – French canals and bridges were privately financed in the seventeenth century, as were the railways in the UK in the nineteenth century. Currently, privately financed projects throughout Europe have included toll roads in France and Spain, and power stations in Italy, Spain and Portugal. There is particular emphasis on reducing public expenditure with regard to satisfying the criteria for monetary union under the Maastrict Agreement. However, it must be noted that the targets set under Maastrict are subject to the European Union's standard measures. Therefore reducing the public sector borrowing requirement (PSBR) will only go towards meeting this target if the PFI project is wholly financed by the private sector.

The pressure on public finance is not restricted to Europe. The need to cut, or restrict, public expenditure is a global objective being pursued in most countries. There is growing interest in the use of private finance all over the world, especially in the emerging economies in Asia. Merna and Owen (1998) state that an International Financing Review (IFR), *Project Finance International,* has reported a 100% increase in infrastructure projects in Asia and Australasia.

How to implement PFI projects

There are different types of projects that fall within the realms of a PFI, the emphasis being placed on deals rather than rules. The UK Treasury has described three strategies: financially free-standing projects, joint ventures and services sold.

Financially free-standing projects

These ventures are financed and managed entirely by the private sector. The project is financed by the payments made by the end users. This type of project does not require a value for money (VFM) test, only government approval. Examples are the second Severn Crossing and the Skye Bridge. It is envisaged that some road schemes will fall into this category in the future.

Joint ventures

These are public and private sector partnerships where the private sector retains control. The public sector is involved, as it perceives some public benefit that cannot be accounted for in strict monetary ways, such as a reduction in traffic congestion in certain areas. This type of project does require a VFM test and will conform to the following criteria:

- the private sector partner is selected through competition
- joint venture control is held by the private sector
- there is a clear definition of the government contribution and its limitations
- there is a clear agreement about risk and reward allocation, defined and agreed in advance, ensuring that the private sector returns are genuinely running some risk.

Government contributions can take various forms, including equity, concessionary loans and asset transfer, but equity stakes are not controlling. Typically, governments contribute through aiding in the initial planning or through a grant subsidy. Local authorities can also contribute, and this is aided by the relaxation of the capital receipt regulations placed on them.

Services sold

The services sold can take various forms. An example is a private firm supplying the National Health Service with kidney dialysis services, a clinical waste incinerator or a hospital information support system. Prison provision is a particularly good example of this, with consortia providing accommodation and prisoner services for HM Prison Service. There is an element of leaseback in some projects. In order to avoid this showing on the spending allocation of the public bodies it is necessary to include a service incorporated with the facility lease. For example, HM Prison Service does not lease the prison accommodation, but pays for a complete service provided to the inmates. This is an important distinction to make.

The UK Private Finance Panel

The Private Finance Panel (PFP), an independent body set up by the Treasury, was convened in 1993, a year after Norman Lamont announced the PFI. Its initial chairman, Sir Alastair Morton, was charged with the responsibility of initiating projects whereby private finance could be introduced for the provision of public services. The chairman led a panel of eight unpaid non-executives who shared the common ground of interest in and commitment to making the PFI work. The panel was advised by 24 executive members (from January 1996), who acted as the problem solvers and prompts of action. The executive panel was the focus for the participants of a PFI to discuss problems and seek advice and solutions. The PFP offered an advisory and guidance facility for those involved in or wanting to become involved in PFI schemes. The PFP was originally used as an intermediary to move the policy forward and to try to unblock the failure of communication coupled with failures in attitude.

The role of the PFP constantly evolved, and in April 1996 new roles were defined in an attempt to smooth the procurement process. These were:

- to act as a central repository of information about people offering advisory roles in the PFI
- to produce case studies of previous PFI projects in order to improve the learning curve and prevent similar mistakes from being repeated

- to provide a training course at the Civil Service college in association with Price Waterhouse, allowing for specific training of civil servants and the enrolment of the private sector.

Many commentators, primarily clients and financiers, believed that there was a lack of confidence in the PFP. They believed that the PFP was not an effective vehicle for driving the PFI forward. However, there was, and still is, a great need for such a drive, in order to capitalise on and pursue the tentative interests of leading industrial and commercial groups.

Many criticise the efforts of the PFP in trying to streamline and speed up the PFI. They indicate that the guidance documents produced by the PFP were too basic and did not cover the issues in sufficient detail, with specific reference to the attempts to develop a standard form of contract.

The Public–Private Partnership Programme Ltd

The Public–Private Partnership Programme Ltd (4Ps) was set up in April 1996, enjoying all party support, with an initial estimated life of 3 years. Essentially the 4Ps was modelled on the Private Finance Panel Executive (PFPE) and Private Finance Units (PFU) in government departments, but with specific emphasis on aiding local authorities. It came into effect on the back of local authorities continually faced with increased constraints on infrastructure provision, with the aim of helping by encouraging greater investment in local services through partnerships between the public and private sectors.

The stated objectives of the programme can be outlined as follows:

- to lobby government to improve the process of implementing the PFI when regulation proves restrictive
- to assist the Audit Commission in providing clear guidance on PFI and partnership schemes to district auditors
- to assist in delivering Pathfinder projects in housing, education and transport
- to compile a database of proposed and recently completed projects for future reference, thus preventing continual re-invention of the wheel.

Both the 4Ps and the PFP opened databases allowing public and private sector organisations to register their details. This enabled interested parties to obtain information on project progress and company experience.

Impacts of the PFI

The PFI faced many teething problems and risks. Risk allocation was a major source of problems for the early PFI projects. Risk can be allocated in two ways: through payment mechanism or through specific contract terms. As the public sector has no need to own the facilities it uses, the public sector takes the role of setting specifications of the services to be provided by the assets. This is a great change from the past. Many private operators felt that they had been expected to shoulder too much risk. This prompted a publication by the PFP (1996) entitled *Risk and Reward in PFI Contracts*.

Some concessionaires are concerned that risk transfer from the public sector to the private sector is too great. Some consortia feel that local authorities do not properly understand the spirit of the PFI and try to transfer all the risks to the private sector in order to fulfil their 'off balance sheet' accounting objectives. Due to a lack of understanding about the PFI process, some schemes can suffer from increased risk due to inadequate scoping of the project. In some cases, consortia have felt that they are expected to run risks that are beyond their control, such as legislative risks, but which may be influenced by the public partner. However, the counter argument to this is, what kind of protection does a wholly private sector company receive in the normal business environment? In most cases this is a common business risk borne by most of industry. The question then arises of whether the PFI special project vehicles (SPVs) should lay claim to protection from this aspect of risk.

With PFI sanctioned projects there appears to be little scope to deal with work redefinition (there does not appear to be provision for this within the contract) and hence a greater risk is run by consortia. This is different to other foreign countries, particularly France, where public–private works contracts have clauses that allow for re-scoping of work. Built-in mechanisms allow for the contractor or operator to be fairly compensated for this change.

Even when local authorities are willing to shoulder greater risk through underwriting projects, it appears that recent judgements could prevent this occurring. Credit Suisse action against the Borough Council of Allerdale and Waltham Forest London Borough Council saw the Court of Appeal rule that the local authority in question was not legally able to provide such a guarantee on the project. The piece of legislation in question is Section 111 of the Local Government Act 1972. In its interpretation, the Court of Appeal decided that the local authority had acted outside its statutory function and therefore was not obliged to honour any guarantees it had made to the banks. This has implications on other areas of PFIs, where government bodies may not have the power to form their own development companies, and thus to finance the development the company would have to guarantee any loans it sought to acquire. Assurances were sought by potential consortiums regarding the impact of a trust becoming insolvent and the implications on the service contract between the two parties.

Bidding in PFI projects

If a PFI project requires some investment from the public sector, i.e. it is a public–private partnership, the project has to satisfy a VFM test. The VFM test is used to determine the following three factors:

- whether the project will proceed at all
- if it is decided that the project will proceed, whether the project will proceed on a PFI or a traditional basis
- which PFI supplier is chosen to provide the service.

A secondary application for the VFM test is in achieving the proposed benefits of the PFI. Some additional benefits that can follow from applying a VFM test are that it can:

- ensure fitness for purpose, e.g. avoid overdesign and lead to integration of operational needs to achieve optimum efficiency
- increase the efficiency of construction and operation by allocating expertise using new technology or by applying more effective business processes
- identify potential economies of scale, e.g. by using the asset to service a third party client (i.e. to extend the customer base)
- lead to a design of the asset that maximises the residual value or allows for a change of use at end of the concession period.

This VFM process impacts on the procurement process and the time taken to begin the construction phase. Hence a commonly stated problem with PFI procurement is the time of tendering and the VFM process itself. Many bidding organisations complain that, in addition to the problem of time, there is also a significant cost element involved. This is due to the way in which the competition is administered, with the awarding body allowing perhaps four consortia to bid and then selecting two for the final stage.

This process involves a great deal of expense on behalf of the bidders, with one ultimately losing. However, under current rules this is how the competition must be run. These problems have been quoted as the reason behind several high-profile withdrawals, with large firms such as Bovis deciding against any further design–build–finance–operate (DBFO) road bids, and Laing pulling out of several hospital projects and some preferred bidders being selected by default as the competition withdrew. Critics argue that this signals the demise of the PFI and that industry is turning its back on the initiative. This would appear not to be the case, as consortia are pulling out to bid for alternative PFI schemes rather than abandoning the idea altogether.

Many companies found themselves bidding in unfamiliar business environments and decided to concentrate on their specialist activities rather than to diversify into completely new areas. This makes a great deal of sense when considering the cost and the resources needed to submit a good bid. The pathfinder prison report states that the total time from advertising in the *Official Journal of the European Community* to closure was 25 months, with 17 months from the invitation to tender (ITT) to commencement. The aim is to reduce the time from ITT to contract award to 9 months.

Industry commentators warn that the variety and number of PFI schemes in the pipeline could be so great that there is an implication that competition could be eroded, with projects chasing consortiums. This has the potential to allow the private sector to dictate terms that it would bid for, hence allowing the private sector to pick and choose profitable projects, leaving the less attractive ones. In this case service provision would be made on the basis of economic viability rather than need. This problem could be diminished if there was an influx of overseas private operators into the PFI market. At present, the threat of increased competition is slow, but it would surely benefit the public sector. Such companies are experienced in operations and have a track record in service provision, and thus may stimulate greater competition.

Contractors' reactions to PFI projects

It was initially assumed that contractors would be the main bodies behind PFI consortia, but there is a problem of whether they are the best people to operate a facility once it has been built. In general, large contracting organisations have the operational and maintenance expertise to run a PFI facility. The motives for entry into the PFI ought to be considered. It can be argued that in a time of limited opportunities for conventional projects, contractors are merely chasing work, but with conventional margins running at 0–3%, the PFI offers a good opportunity to diversify their work and realise greater margins. A major failing with UK contractors is their inability to contribute equity to the finance package. This could change with the acquisition of large UK contractors by financially strong overseas counterparts. At present, balance sheets for many

UK contractors are not as healthy as, say, their French counterparts, and this restricts their involvement in PFI schemes as they have less capital to invest. However, with the influx of foreign investment in the form of large stock acquisitions of major UK construction companies (such as Malaysia's Intria Berhad into Costain, Kvaerner into Trafalgar House and Bouygues into Tarmac), this situation is likely to change. It seems that many Middle Eastern and Asian firms are waiting to tap into the UK's civil engineering expertise.

The contractor involvement in a PFI is somewhat dependent on the type of project in question. It is generally accepted that financially free-standing static infrastructure projects, such as toll bridges and tunnels, are ideally undertaken by contractor-led consortiums, as this does not represent a shift away from their traditional work, with the exception of managing the revenue stream. However, when considering non-clinical health care provision, it can be argued that a contractor is not the best person to manage that service.

Confusion and ignorance were cited as early reactions to the PFI procurement strategy. The PFI is described as the biggest challenge to the construction industry, and some commentators predicted that 20% of all new buildings would be PFI procured prior to 2000.

This poses an obvious threat to UK contractors, as limited experience and knowledge will not improve chances of gaining work, although many contractors are perfectly happy with build–own–operate–transfer (BOOT) strategies, which share some common ground with the PFI. Unless these problems are modified, the UK industry may expect competitors from the USA and continental Europe to enter the market.

Obvious opportunities to the industry are the healthier margins achieved in PFIs than in the more traditional strategies (Merna and Owen, 1998, citing 1995 figures for the top ten companies).

Operators, rather than construction companies, now lead many PFI projects. Although many PFI projects have been dominated by construction-led promoters, the trend now seems to be towards operator-led promoters, as the economies of the project are achieved over the life cycle of the project, where the constructor is unlikely to have the required operator skills.

Consultants' reactions to PFI projects

Some consultants are particularly worried about the growth of PFI funding, as it is a radical departure from traditional public works procurement. The potential is that if the bidding consortia are led by the large contracting organisations they will possibly already encompass a good design capability, and this could signal a reduced requirement for consultancy services. Consultancies can counter this by joining the consortia themselves, but this would require capital, which many cannot provide. Consultants are increasingly modifying their services, with a plethora of PFI technical advisors now appearing. They join the ranks of the public bodies' advisory teams, incorporating financial, legal and technical disciplines. Although consultants appear to be able to command premium fees for this advisory work, there are two major disadvantages:

- Many public clients demand bidding in the form of fees incorporating a success element. This is troublesome, as many hours of work are undertaken, in some cases for little or no payment. Consultants are running this risk, particularly on hospital projects, as the success element would be considerable, but to date elusive.
- The advisory team is relatively small in comparison to a design team and is possibly retained for a shorter period.

The advantage of acting in an advisory role is that there is scope to secure downstream design business. Hence, gaining entry into the client's team at the point of facility or service definition is extremely important. Consultants face a similar problem of inexperience as contractors, which hampers their chances of gaining advisory positions. With many projects stalled or cancelled it is difficult for public clients to separate good from bad, making them suspicious of anyone offering advisory capabilities.

The nature of the PFI changes the way in which consultants respond to clients. Instead of a design to meet a technical specification, consultants have to react to an output or service specification. To many firms this is unfamiliar territory and one that requires a careful approach. Clients are asking whether design consultants have the strength and depth of skill base to cope with this change in demand.

The VFM paradox in PFI projects

A paradox that may occur in PFI projects is that in trying to achieve VFM through the transference of risks to consortia, actual VFM may deteriorate. This is of particular concern because one of the cornerstones behind the PFI is achieving VFM. It is an important fact to bear in mind that the private sector borrows at 6–9% points above gilt rates, and hence the private sector must more than make up for this in efficiency gains if VFM and profitability are to be achieved. It is because of the high cost of financing that some critics argue that the PFI is fundamentally flawed. An interesting argument is illustrated in Figure 8.1.

The figure shows that as the risk transfer increases, the greater is the level of risk that is borne by the consortium. This means that a greater amount of equity is required, as banks would be less willing to finance such deals. Hence higher equity means higher financing costs, as demonstrated by the curve. As the project will be paid for by taxpayers in the long run, there is an argument that VFM may not be achieved and the public sector will pick up the burden. The relationship between VFM and risk transfer is shown in Figure 8.2.

In projects where there is a large transfer of risk, the private sector will not be reluctant to bid for the project as was the case in early PFI projects, but will price the risk. This means that project costs grow and possibly violate the VFM criteria. Hence it is important that risk transfer is not excessive.

In addition to this paradox, there are inefficiencies that must be addressed. These include the cost of compulsory evaluation of projects to see if the PFI is the suitable strategy, and the cost of assessment if the project does meet the VFM criteria. These costs are due in part to the team of PFI advisers consisting of technical, financial and legal experts. In addition to this there is the large cost to industry of bidding for PFI projects, especially when the scheme is cancelled later on.

The public sector comparator

In order to demonstrate VFM, it is necessary to undertake a study of a public sector comparator (PSC). In the majority of cases there will be an historic publicly funded project that can be compared to the PFI approach. From this it will be possible to assess whether VFM is achievable. In a minority of cases there will be no public equivalents, but good practice suggests that a realistic solution should be

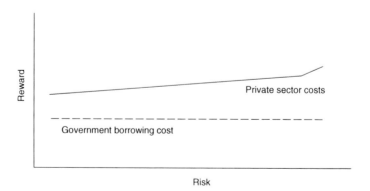

Figure 8.1. The cost of risk transfer

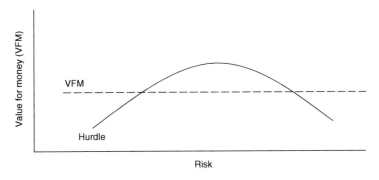

Figure 8.2. Risk and VFM

developed. In the VFM appraisal, a do-nothing option should be included. This is important, since this alternative can be costed alongside the proactive approach. In the case where no public alternative is available a proxy method can be achieved by constructing an artificial comparator. In the design–construct–manage–finance (DCMF) prisons, Bridgend and Fazakerley, this was done by modelling a series of hypothetical contracts, and the do-nothing scenario involved looking at the cost of housing prisoners in police cells.

The PSC will focus on the life-cycle costs of the project in order to achieve the stated output requirements. An important section of the study needs to include the value of risk, which the public was previously responsible for and now wishes to be passed on to the private sector. This process is used to demonstrate and satisfy the auditors that the decision to undertake the project is sensible and justifiable. In the case where a PSC is not necessary, VFM is achieved through the use of competition.

Potential PSC distortions

It should be borne in mind that the PSC easily projects an optimistic level of costs. This is down to the ways in which the public and private sector deal with, and are accountable for, commercial risk. The public sector is likely to underprice the risk it bears, as it is not accustomed to bearing the full financial risks that a private company usually bears. It is, however, important to try to accurately price the risks involved at the market rate.

The PSC flowchart from the PFP publication *Private Opportunity Public Benefit* (1995) (also in Merna and Owen, 1998) shows the relationship between VFM and the PSC. The flowchart shown in Figure 8.3 identifies the process of how the PSC is used and identifies whether a VFM test is required.

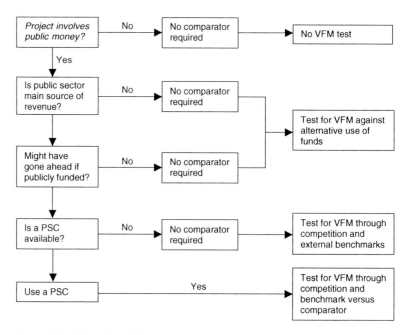

Figure 8.3. How the public sector comparator is used

Output specification

One of the guiding philosophies of the PFI is that the public bodies allow the private sector the freedom to determine how it is to provide a service to meet the relevant specification. In order to give the private sector this freedom, the client has to be very careful in the way he expresses his requirements. This is achieved through the drafting of an output specification. This puts public bodies in an unusual situation, as they have to be the authors of a service provision document rather than an asset provision document. Although this may seem straightforward, it is sometimes difficult to separate the service from the facility, as this leads to preconceptions of how a service should be run and from what type of facility. Such presumptions should be avoided at all costs, as they remove the freedom of the private sector to innovate and produce alternative solutions.

In May 1997 the National Audit Office criticised the Scottish Office for letting the promoter of the Skye Bridge levy tolls at a level of 9% higher than envisaged. Lack of competition was cited as the reason for the Scottish Office not getting the best terms for the concession when negotiating with the promoter. The National Audit Office stated that best VFM was not achieved since the financial negotiations were not completed until the contract was awarded.

Facilities provided sectors

The PFPE identified many areas of the public sector that they believed were suitable for a PFI. The UK Treasury categorises them as follows:

- defence
- agriculture
- trade and industry
- education and employment
- transport
- environment
- home office
- legal departments
- national heritage
- health
- social security
- Scotland
- Northern Ireland.

There has been massive interest in the provision of services for the health sector, which has attracted much negative publicity and the conspicuous absence of signed deals. According to the PFP, there had been 47 signed contracts for a PFI service provision for the Department of Health by mid-1997, ranging from clinical waste incineration to the provision of information systems and office accommodation. Those looking for new hospital provision were disappointed, as only two deals had been signed by 1998, Norwich & Norfolk and Dartford & Gravesend being valued at £170 million and £115 million, respectively. Both are subject to financing, indicating that the banks are less than confident about the projects. The figures quoted in the 1997 autumn edition of *Private Finance Quarterly* outlined some £2600 million of proposed health service expenditure. This figure demonstrates the level of interest by trusts to see new PFI hospitals, but the process appears to fall down at the financing stage. The Dartford & Gravesend project leapfrogged the Norwich hospital, with construction starting in April 1997. This may be because a detailed contractual framework was drawn up prior to inviting bids, which improved the process. The former Chair of the Private Finance Panel, Douglas Hogg, suggested possible early involvement in the project by financiers could ease the process. Although this may possibly restrict competition it would define minimum requirements, and hence prevent the progression of unfinanceable projects.

A sticking point with high-value PFI hospitals is that trusts do not appear to be realistic in their requirements, and consortiums responding to output specifications spend large sums of money preparing bids which ultimately are not affordable. This problem arises because of the intensive level of borrowing required by the consortium in the early stages of a project and the subsequent debt servicing. These costs are passed on to the trust, through charges levied for services during the operation period. To rectify this problem a mechanism was proposed for the Swindon & Marlborough Trust that would allow for the transfer of money from the NHS capital budget to the PFI project. This would help overcome the initial high costs and would then be repaid at the end of the concession period at an interest rate of 6%.

The prisons sector has been the beneficiary of modest PFI success, with three privately financed prisons constructed and several others in the pipeline. The DCMF prisons were a progression from privately operated facilities. These existed in three prisons across the UK, including HM Prison Doncaster and Wolds Remand Prison. These prisons are still operating successfully, showing significant cost benefits in comparisons with existing best performers in the public sector. Alongside this was the contracting out of prisoner escort services, which are provided by Group 4. This initiative was to eliminate inefficiencies apparent in the current facilities. The natural course of action follows that, in order to expand the capacity of prisons through building more prisons, the potential private sector operators should provide input into the design, construction and financing of the asset. The existing links within HM Prison Service and experience with private operators meant that this service was well suited to PFI treatment. There was an existing culture within HM Prison service to work alongside private organisations.

In addition to contractors turning into operators and bidding for projects, there was the arrival of the US operator Wackenhut Correction Facilities, who joined the bid as consortium members. This existing track record and the existing privately operated prisons in the UK meant there was already in place a valuable benchmark against which to evaluate the bids. The success of the pathfinder projects was aided by the government's desire to see the PFI realised, and the help they provided reflected this. In the Fazakerley and Bridgend prisons, the government acted as insurers as a last resort if the operators could not obtain commercial insurance. The follow-up projects, particularly in Lowdham Grange, hit problems due to the government's refusal to repeat this. This potentially terminated the concession agreement with Premier prisons unless their record in Doncaster could reassure the insurers as to their record on riot and disturbances. It was felt by HM Prison Service that the success of Fazakerley and Bridgend meant they could transfer greater risks in forthcoming projects. They felt that commercial insurance would provide incentives to operators to maintain an exemplary record when it came to discipline, as no claims bonuses could be earned.

The PFI under the new Labour government in the UK

In the early 1980s policies to involve the private sector in the provision of services had been ardently opposed by the Labour Party, which at that time was in opposition. However, the Labour Party had gone through great internal reforms and the government elected on 1 May 1997 quickly moved to confirm its support for the PFI.

On 8 May 1997 the new Labour government announced that a review of the PFI was to be undertaken by Sir Malcolm Bates and that the requirement for universal testing was to be removed. Universal testing was the requirement for government departments to produce models for the private financing of all projects, even where this was obviously not suitable. The review made 29 recommendations, including the creation of the Treasury Taskforce to help central government departments and agencies road test significant projects for their commercial viability. The Treasury Taskforce replaced the PFP and was intended to have a 2-year lifespan, which ended in late summer 1999.

A second review of the PFI and public–private partnership (PPP), also by Sir Malcolm Bates, was announced in November 1998 in advance of the expiry of the Treasury Taskforce mandate. The report was published in March 1999 and made 32 recommendations, including considerations concerning the future of the Taskforce. The report considered two possible bodies to succeed the Taskforce:

- a public sector, free-standing agency
- a new PPP.

Bates stated that a decision to go ahead with the latter option would require the development of a business case, identifying the likely risks and rewards.

The recommendation of the second Bates' review was that a private sector body, with a significant public sector stake, should be set up. This would be called Partnerships UK (PUK). PUK would provide development funding, where existing forms of private finance were not available, and its capital base would come from both the private and public sectors. It was intended that PUK would be funded increasingly by private sector loans and equity capital.

It was reported that between the general election in May 1997 and June 1999, 140 deals were signed worth £4.7 billion. In March 2000 the government announced a £20 billion expansion of the PPP programme over the following 3 years, including investment in London Underground, improvements to the air traffic control infrastructure, education, health and transport projects.

At the time of editing (November 2001), two of the above projects, London Underground and air traffic control, have received a lot of press coverage. The former is the subject of possible legal action by the Mayor of London against the government on the grounds of quality should a PPP be adopted, rather than a bond issue proposed by the Mayor himself. The latter became a PPP on 27 March 2001, with the public sector, a consortium of airline operators and employees taking 49%, 46% and 5% stakes, respectively. The sale realised £800 million and a proposed investment of over £1 billion during future years of operation.

The construction of new facilities and the refurbishment of existing educational facilities procured through the PFI can be seen from the following example. A consortium including the engineering and construction group Balfour Beatty were recently awarded a £153 million contract to build nine schools in Stoke-on-Trent by the city council. The 5-year contract is believed to be the big-gest private finance project for schools in the UK. Transform Schools, a joint venture between Balfour Beatty and a private equity company, Innisfree, will build the new schools and refurbish 122 existing schools in Stoke. It will also provide cost-cutting energy management and ongoing repairs and maintenance to Stoke's schools.

However, private finance in the UK has so far been dominated by transportation projects. Major projects have included a £320 million rail link to Heathrow Airport, the £7 billion Channel Tunnel Rail Link, £250 million to build and maintain a new air traffic control centre in Scotland and over £500 million to build and operate trunk roads.

Public–private partnerships

Increasingly, projects that principals wish to realise are not sufficiently robust to be procured by totally private finance funding. It is only some form of public sector involvement that can close the gap between commercial financial analysis and social cost–benefit analysis. The public sector has four main mechanisms for participation, and these operate within a strict hierarchy (Figure 8.4). Essentially, and in sympathy with the concept of encouraging the private sector

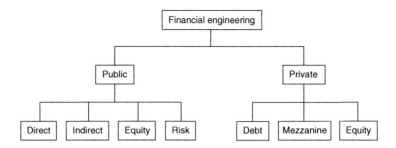

Figure 8.4. Public–private financial engineering

to take responsibility for project risk, the public sector wishes to make the minimum contribution sufficient to reduce the investment risk to levels acceptable to the private sector. Typically, there are three types of partnerships, these being shared, dominant and independent, which are suitable for certain types of strategy and schemes, each having its benefits and disbenefits.

As the first stage, the public sector can consider additional risk sharing. Risk is already shared through the concession agreement, using the basic principle that the party best able to control or manage a risk should take responsibility for that risk. For risks that cannot be controlled by either party it is usual for the public sector principal to take responsibility, because the risk will only have to be paid for should it occur, rather than paying a premium to a contractor to cover the possibility of it occurring. If the public sector principal assumes responsibility for some additional risk the robustness of the project cash flow may be enhanced, and hence investors and lenders may be attracted at an acceptable rate of interest.

Once the risk sharing option has been exhausted and further responsibility for risk is unacceptable to the public sector, the option of additional equity can be considered. Often this might take the form of offering the promoter the use, the revenue and the operation and maintenance of existing facilities. This stream of revenue at the critical stage of the project can be valuable in closing the gap between debt finance and the equity provision that the promoter is able to raise.

An alternative mechanism was adopted for the new Athens Airport at Spata. When originally proposed as a private sector project, the debt financiers required equity funding in excess of the levels that any of the tendering consortia were prepared to offer. The Greek government then suggested a deferral of the project, and during this period a hypothecated tax on all airline tickets was raised. This funding was then utilised as a second source of equity to bridge the gap between the level required by the lenders and the equity raised by the promoter.

Typical indirect options include tax holidays, grace periods and soft loans. A tax holiday is a predetermined period of time over which the promoter will not be liable for tax on the concession. Conventionally, each project is treated as a ring-fenced investment and, in the UK, is liable to be taxed on its operating profits in the same way as any other commercial enterprise. In a number of UK infrastructure projects, a tax holiday of 5–10 years is likely to be most beneficial. Analysis shows that in excess of this period the returns diminish.

Grace periods and *soft loans* can be used separately or jointly to assist the financing of the project. A grace period is a fixed length of time before which a loan has to be repaid. The concession contract has the maximum capital lock-up towards the end of the construction phase, and if a loan can be effectively extended without interest this can be extremely beneficial to the cash flow of the concession. Since the construction period for most infrastructure projects is 2–3 years, a grace period of 5–7 years is of most benefit. Once the construction of the infrastructure project is completed, the initial debt financing that carries high interest rates comparable with the risks associated with construction can be repaid and the debt refinanced at lower rates of interest commensurate with the known risks of operating similar infrastructure projects.

A soft loan is a loan offered at rates below the commercial market rates. The public sector loses some of the return that could have been gained but assists the promoter by reducing the debt burden. The range of these options can be considerable. An example is to consider the financial implications for two concession consortia, one with a public loan at the commercial rate over 12 years and another with the same loan but with a 10-year period of grace and interest at a quarter of the commercial rate.

Finally, if no other course of action is available, the public sector can provide direct financial investment. This may be necessary on projects perceived to be high risk or projects with a marginal cash flow. Examples are the M1 Vienna to Budapest toll road in Hungary where the National Bank of Hungary, the Nemzeti Bank, made a contribution of 40% of the project value in Hungarian Florints. Similarly, most light rail transit schemes in the UK and in many other countries are financially non-viable without a major financial grant from the public sector, often between 75% and 90% of the capital cost. Public sector investment is often justified in projects with a high socio-economic rate of return but a lower financial rate of return.

The structure of project finance

Over the last few years banks have become more accustomed to lending on a *non-recourse* basis. This means that the loan is secured purely on the strength of revenues generated by a facility. Conversely, this is faced with a move by developers away from syndicated loan sources, as they try to achieve a broader portfolio of funding solutions. It is hoped that, because of the current globally low interest rates, investors will achieve higher returns on bond issues. These issues are becoming more popular, with two examples highlighted: the Docklands Light Railway link to Lewisham and the Road Management Group (RMG). The RMG issue was used to fund the Amec and Alfred McAlpine led consortium to undertake the DBFO A1(M) and A419/A417 road projects. This was undertaken using a 25-year bond issue. It was felt that, due to the low levels of risk borne by the consortium, it would be straightforward to market the bonds. The Lewisham scheme was managed by BZW and the issue was based on the well-received issue by RMG, which was for 24 years. Each of these schemes has raised £165 million from issuing bonds. The *Financial Times* reported that there is an emergence of dedicated holding companies whose specialisation is the private financing of previously public projects. These companies can then be floated on the stock exchange. Examples of these are Consolidated Electric Power Asia (CEPA), Southern Company and Central South West Corporation (CSW) from the USA.

The use of bonds has an upside and a downside. With banks reluctant to back a concession period longer than 10 years, bonds suit infrastructure projects particularly well. Bonds also give investors the flexibility to buy and sell publicly listed bonds as and when they desire. Bond financing proves attractive to certain types of institutional investors, such as pension funds, where the managers often look toward investments that match their long-term liabilities. Bonds provide such investment as they often mature for periods of up to 30 years.

With bond financing there is a problem for the consortium on the level of funding to seek and the inflexibility thereafter. The problem arises when a cost overrun occurs and there is the issue of how these extra costs will be met. Also, the consortium does not require all the finance up front, which is a disadvantage of a bond issue. Coupled with this is the problem that projects that come in under budget are liable to pay dividends on unused capital. Unpredictable markets can slam shut an opportunity to raise finance purely on the basis of a change in attitude, totally unrelated to a project's viability. For these reasons it is advocated that bond financing should be complementary to bank lending rather than a replacement for it. This is because most banks have a tendency to have a greater input into a project, and with a hands-on approach.

Summary

It is clear that privately financed public sector services are here to stay. With the general principles of privately financed infrastructure projects enjoying support in many countries, it is necessary for the construction and related industries to adapt. However, it would appear that the most successful projects to date have been of the financially free-standing variety. This fact questions whether the lack of PFI development in certain sectors is due to industrial inertia or a failure at government level. Certainly it would appear that there is widespread frustration at the lack of progress and the below target number of projects financially closed. Perhaps there has not been enough time for dedicated service providers to develop and establish their skills, with an over-reliance on construction-orientated organisations to progress the initiative. At governmental level, gains could possibly be achieved through a more prescriptive approach, especially with regard to local authority PFI development. Local authorities in particular have a generally ambivalent attitude towards the initiative, with many ignorant of the whole concept. Those wanting to tap into the benefits are finding that they do not have the knowledge and experience to develop their plans alone, and hence advisory teams abound. Some governments are deeply suspicious of the PFI and are not in-line conceptually with such issues as asset transfer.

More and more civil engineers are now involved in PFI projects. In many cases they provide the traditional services of design and supervision. It is important that engineers understand the philosophy of the PFI and realise that revenue generation is paramount to commercial success.

Bibliography

Merna, T. and Dubey, R. *Financial Engineering in the Procurement of Projects*. Asia Law and Practice, Hong Kong, 1998.

Merna, T. and Owen, G. *Understanding the Private Finance Initiative*. Asia Law and Practice, Hong Kong, 1998.

Owen, G. and Merna, A. The private finance initiative. *Engineering Construction and Architectural Management*, special issue 3 (1997).

Private Finance Panel. *Risk and Reward in PFI Contracts*. HMSO, London, 1996.

Private Finance Panel. *Private Opportunity Public Benefit*. HMSO, London, 1995.

CHAPTER NINE

Challenges and opportunities for infrastructure development in developing countries

Introduction

The development of infrastructure projects in developing countries is a challenging exercise. In many developing countries the need for infrastructure services is enormous and exceeds the existing financial resources. In addition, other resources, such as materials and equipment, necessary for the development and delivery of decent infrastructure services have to be imported from industrialised countries at great cost. With the existing global trading environment being largely unfavourable to developing countries, financing of infrastructure projects will remain a challenging task for most developing countries. The high levels of debt in most developing countries mean that new methods of financing infrastructure projects have to be sought.

What are developing countries?

Countries are often classified according to their economies for analytical purposes. The main criterion for classifying economies is gross national product (GNP) per capita. Every economy is classified as low-income, middle-income (subdivided further into lower middle and upper middle) or high-income as follows:

- *low-income economies* are those with a GNP per capita of US $675 or less in 1992
- *middle-income economies* have a GNP per capita of more than US $675 but less than US $8356 in 1992; a further division at GNP per capita of US $2695 in 1992 is made between *lower-middle income economies* and *upper-middle income economies*
- *high-income economies* are those with a GNP per capita of US $8356 or more in 1992.

Low-income and middle-income economies are sometimes referred to as *developing economies*. The use of the term is meant to be for convenience and not intended to mean that all economies in the group are experiencing similar development or that other economies have reached a preferred or final stage of development. Classification by income does not necessarily reflect development status. The term 'countries' is widely used to refer to 'economies' in the context of this classification. The term 'developing countries' refers to 'developing economies', i.e. countries with low-income or middle-income economies.

Developing countries are also commonly known as *less developed countries*. Wells (1973) defines less developed countries as:

> ... those where incomes per head are relatively low, where only limited supplies of capital are available, and where a high proportion of the labour force is engaged in primary production. The most important feature of these countries is their poverty.

Economies in developing countries

Economic problems are evident in most developing countries and can be seen in many forms, which include:

- deteriorating or collapsed infrastructure, such as roads, water systems, power supply, hospitals, schools and municipal services
- rampant unemployment and poverty
- low and falling levels of per capita income, exacerbated by high population growth rates
- unsustainable domestic and external debt perpetuated by unmanageable budget deficits arising from high public expenditures
- excessive donor dependence to meet public expenditures.

The root causes of the present economic problems in most developing countries are:

- A heavy reliance on primary commodities to earn foreign exchange.
- An unfavourable international trading environment, largely skewed in favour of industrialised countries.
- High rates of population growth, accompanied by falling economic growth rates.
- Most developing countries were drawn into the East–West conflict, and many opted for some degree of the socialist economic model of the East in which the state controlled and managed the economy. That model has since been proved to be a failure. Those countries that stuck with the West did not fully reform their economies, but instead remained perpetually dependent on western financial and technical assistance.

- The leaders of many developing countries devoted most of their energies to political issues and gave lip service to the equally important economic matters.
- Excessive economic controls were introduced after independence, with disastrous consequences.
- The leadership and institutions in many developing countries were poor, resulting in greed, corruption and ethnic problems, all of which had negative impacts on the economies.

The challenge of infrastructure development

Developing countries face an enormous challenge in meeting the infrastructure requirements of the growing population. The challenge of financing and management of infrastructure projects is evident from demographic figures and trends. Poverty, lack of capital and national debt further complicate the challenge facing many developing countries.

Infrastructure services in many developing countries are inadequate, poorly managed and inefficient. The rapid population growth, especially in the urban areas, adds strain on existing infrastructure services and further compounds the problem. These two problems, when combined with the prevailing poverty, are stifling economic growth and reducing an already low standard of living in most developing countries.

It is estimated that of the 6 billion people living on this planet, less than 1 billion live in what the World Bank categorises as high-income countries, while about 3.5 billion live in low-income countries. The United Nations estimated the world population as 5.8 billion in 1998, of which 79% lived in developing countries and 21% in developed countries. The world population is projected to be over 7 billion in 2015, with 82% living in developing countries and 18% in developed countries. Of the 4.6 billion people in developing countries in 1998, 39% lived in urban areas. The urban population in developing countries is rising rapidly and is estimated to constitute over 49% of the total population (projected at 5.8 billion) by 2015.

The United Nations estimates that over 1 billion people living in developing countries lack access to safe water and over 2.4 billion lack basic sanitation. In addition to this lack of basic water and sanitation facilities, many people live in substandard housing; transportation and communication links are poor or non-existent; and fuel for heating, lighting and cooking is in short supply. A high proportion of the population in developing countries does not have access to decent levels of infrastructure services, and this proportion is increasing. The World Bank reported recently that African cities are growing exponentially at an average of over 5% per annum, and over 9% per annum in some cities, while India's population is now over 1 billion. The costs of providing infrastructure services are prohibitive.

Debt is another problem facing many developing countries. The UK Department for International Development has stated that, by trying to pay back the money they owe to richer countries, developing countries no longer have enough resources to spend on vital infrastructure services, including health, education, water and sanitation. Levels of debt are high and official development assistance reduced by 24% between 1992 and 1998.

Poverty and lack of capital in developing countries complicates the development of infrastructure. Data from the United Nations show that over 1.2 billion people in developing countries are income poor. Over two-thirds of the world's population live in countries whose income per head is below £100 a year, and together these countries account for less than 13% of world income. At the other extreme, 7% of the world's population living in countries where income per head is over £500 a year enjoy no less than 36% of world income.

Perhaps the most serious handicap facing infrastructure development in developing countries is the grave shortage of capital. Because incomes are low, savings are low, and because savings are low, investment is low. Hence such countries persist in unsophisticated traditional methods of cultivation, their manufacturing industries tend to be technologically inferior to those of the developed countries and their infrastructure systems are inadequate. There is a vicious circle of poverty from which developing countries can hope to break out only by massive development programmes financed from external investments from wealthier countries. The challenge of financing and managing infrastructure projects in developing countries is evident. This challenge can be met by well-conceived and well-managed projects with active participation of both the public and the private sectors. The opportunities for the private sector are evident.

Opportunities for infrastructure development in developing countries

The challenge for infrastructure development in developing countries is enormous at a time when shrinkage of the domestic construction market is causing anxiety in many developed countries. Many firms are looking overseas for work that will take up the slack. There is a broad consensus that infrastructure development in developing countries will increasingly offer significant business opportunities to firms based in developed countries. It is therefore necessary for practitioners in developed countries to understand the nature of infrastructure financing in developing countries.

Considering the above scenario, it is unlikely that governments in developing countries can keep up with the infrastructure requirements of the rising population using the traditional public financing of infrastructure projects. A key objective is for governments to seek the active participation of the private sector in financing and managing infrastructure projects. Merna and Njiru (1998) consider the adoption of innovative methods of financing and managing infrastructure projects as key to meeting the challenge of providing infrastructure to the growing population in developing countries. Many governments in developing countries have realised the need for innovation and are making efforts to create an enabling environment for private sector participation in the financing of infrastructure development projects.

Traditional methods of financing infrastructure in developing countries

Provision of infrastructure has traditionally been the preserve of the public sector. Funds for the development of infrastructure projects are traditionally obtained from general taxation or borrowed from multi-lateral and bilateral agencies. Due to low levels of public finance derived from general taxation, most developing countries rely on borrowing from multi-lateral and bilateral development agencies to finance infrastructure development. Most of the developing countries now have a great debt burden and are spending a large proportion of their meagre finances in meeting debt payments. Whereas these countries used to borrow in order to invest in the development of infrastructure, a large proportion of the funds borrowed is now used to service debts, with little going to finance infrastructure development projects. The level of finance available for borrowing from the traditional sources has reduced in the recent past.

Traditional methods of public financing and management of infrastructure projects have failed to keep pace with the rising demand for infrastructure services in most developing countries. The private sector has, traditionally, only passively participated in the provision of infrastructure as consultants and contractors during the implementation phase of infrastructure development projects that are financed and managed by the public sector. In recent years, many countries have seen the need to look for alternative methods of financing and managing infrastructure projects. The private sector has often been called upon to participate actively in the financing and management of infrastructure projects. Such participation is possible and sustainable only if the objectives of both the public and the private sectors are met, while providing users with quality infrastructure services at a competitive price.

Why developing countries should invest in infrastructure projects

The benefits of infrastructure development are numerous and may be grouped broadly into three categories:

- human welfare and economic development
- reduction of poverty
- improvement of the environment.

These are discussed below.

Infrastructure covers a complex of distinct sectors that together represent a large share of a country's economy. The services associated with infrastructure account for about 7–11% of GDP. Infrastructure typically represents about 20% of total investment and 40–60% of public investment in most developing countries. Public infrastructure investment ranges from 2% to 8% (average 4%) of GDP. The social and economic importance of infrastructure is enormous.

Infrastructure has strong links to growth, poverty alleviation and environmental sustainability. Research on the impact of infrastructure on growth has concluded that the role of infrastructure in growth is substantial, significant and frequently greater than that of investment in other forms of capital.

Infrastructure services, such as power, transport, telecommunications, provision of water and sanitation, and safe disposal of wastes, are central to the activities of households and to economic production. Providing infrastructure services to meet the demands of businesses, households and other users is one of the major challenges of economic development.

Infrastructure represents the wheels of economic activity. Infrastructure services are used in the production process of nearly every sector. Transport is an input for every commodity, while life, let alone economic activity, is impossible without water. Users demand infrastructure services not only for direct consumption but also for raising their productivity by, for instance, reducing the time and effort needed to secure safe water, to bring crops and other products to the market, or to commute to work.

Adequate quantity and reliability of infrastructure are key factors in the ability of countries to participate and compete in international trade, even in traditional commodities. The competition for new export markets is especially dependent on high-quality infrastructure. During the past two decades, increased globalisation of world trade has arisen not only from the liberalisation of trade policies in many countries, but also from major advances in communications, transport and storage technologies. These advances centre on the management of logistics (the combination of purchasing, production and marketing functions) to achieve cost savings in inventory and working capital and to respond more rapidly to customer demand. About two-thirds of production and sales in the Organisation for Economic Co-operation and Development (OECD) countries are processed directly to order, and just-in-time delivery of products has become the norm in many sectors. Because about 60% of their exports are directed to OECD markets, developing countries must meet these standards. Virtually all the improved practices designed to reduce logistics costs, including those in transport, have been based on information technologies using telecommunications infrastructure. Appropriate logistical support provided by efficient transport and telecommunications infrastructure is essential for developing countries wishing to compete in global markets, or to participate in *global sourcing* (the linking of businesses in several countries producing different components for a final product). Diversification of trade is permitted by appropriate logistical support facilitated by efficient infrastructure services. Examples are the recent growth in the computer software industry in India, manufacturing assembly operations in Mexico and horticultural exports from Kenya.

The economic return of infrastructure investment varies not only by sector but also by its design, location and timeliness. The effectiveness of investment, whether it provides the kind of services valued by users (responding to *effective demand*), depends on characteristics such as quality, reliability and quantity. It is essential to match supply to what is demanded. The efficiency with which infrastructure services are provided is also a key to realising potential returns. The availability of infrastructure services valued by users is crucial for the modernisation and diversification of production.

Developing countries that wish to attract foreign investors need to develop their infrastructure in order for investors to invest in the country. Surveys of prospective foreign investors over a wide range of countries show that the quality of infrastructure is an important factor in ranking potential sites for the location of direct investment. The nature of an economy's infrastructure is central to its ability to respond to changes in demand and prices or to take advantage of other resources.

Infrastructure is important for ensuring that growth is consistent with poverty reduction. Access to at least minimal infrastructure services is one of the essential criteria for defining welfare. To a great extent the poor can be identified as those who are unable to consume a basic quantity of clean water and who are subject to unsanitary surroundings, with extremely limited mobility or communications beyond their immediate settlement. As a result they have more health problems and fewer employment opportunities. The burgeoning squatter communities surrounding most cities in developing countries typically lack formal infrastructure facilities.

Different infrastructure sectors have different effects on improving the quality of life and reducing poverty. Access to clean water and sanitation has the most obvious and direct consumption benefits in reducing mortality and morbidity. It also increases the productive capacity of the poor and can affect men and women differently. For example, the poor, and poor women in particular, must commit large shares of their income or time to obtaining water and fuel wood, as well as to carrying crops to market. This time could otherwise be devoted to high-priority domestic duties, such as child care, or to income-earning activities.

Access to transport and irrigation can contribute to higher and more stable incomes, enabling the poor to manage risks. Both transport and irrigation infrastructures have been found to expand opportunities for non-farm employment in rural areas. By raising the productivity of farms and of rural transport, both an increase in the incomes of rural workers and a reduction in food prices for the urban poor can be achieved.

The benefits of transport and communications include the access they provide to other goods and services, especially in cities. Where the poor are concentrated on the periphery of urban areas, as in many developing countries, the cost and availability of public transport become key factors in their ability to obtain employment.

The construction and maintenance of some infrastructure (especially roads and waterworks) can contribute to poverty reduction by providing direct employment. Civil works programmes, which often involve the provision of infrastructure, have also been important in strengthening famine prevention and providing income.

The provision of infrastructure is a result of individual and community effort to modify their physical surroundings or habitat in order to improve their comfort, productivity and protection from the elements of nature, and to conquer distance. Each infrastructure sector (water, power, transport, sanitation, irrigation) raises issues concerning the interaction between man-made structures (and the activities they generate) and the natural environment.

Environment-friendly infrastructure services are essential for improving living standards and offering public health protection. With sufficient care, providing the infrastructure necessary for growth and poverty reduction can be consistent with a concern for natural resources and the global environment. At the same time, well-designed and managed infrastructure can promote the environmental sustainability of human settlements.

The relationship between each infrastructure sector and the environment is complex. The most positive impacts of infrastructure on the environment concern the removal and disposal of liquid and solid wastes. But much depends on how disposal facilities are planned and executed. Under-investment in municipal sewerage relative to water supply in densely populated cities can lead to harmful contamination of water reserves and a reduction of the health benefits from water investments. Provision of sewerage without waste water treatment can lead to severe downstream pollution and public health problems where receiving waters are used as the drinking water supply or for recreation, irrigation and fisheries. Poor management of solid waste complicates urban street drainage and has been linked with the proliferation of disease-bearing mosquitoes in stagnant water.

Funding sources for developing countries

For developing countries to be able to implement their capital investment programmes, such as the development of infrastructure, which is so essential for economic growth, a massive transfer of resources from wealthier countries is necessary. In the past, direct transfer of resources from rich to poor countries by means of grants and loans was considered to be crucial. The current trend is to effect this transfer of resources through foreign direct investment (FDI), and to use less of public sector borrowing from multi-lateral and bilateral agencies. This change has been brought about partly by the high debt levels in many developing countries and partly by the changing political dispensation in the western world. The end of the Cold War has brought new demands and opportunities in Eastern Europe. Demand for capital investments from the traditional sources is now higher than ever.

Aid takes several forms. It may take the form of outright grants, loans and assistance in kind and technical aid. The flow of resources may be in the form of private investment or *government (official) aid*. Government aid in turn is subdivided into *bilateral aid*, that is government to government aid, and *multi-lateral aid*, which is made available to the recipient country through one of the international agencies. In all these cases, a direct transfer of resources occurs-although if the aid consists of loans, repayment eventually involves a transfer in the reverse direction.

After general taxation, the other traditional source of finance for infrastructure projects is international agencies. International agencies often provide technical assistance, grants and loans to the public sector as funding for infrastructure projects in developing countries. Governments repay these loans using public funds obtained through general taxation. The international sources of public funds are multi-lateral and bilateral agencies. These consist of development banks, aid agencies, and charitable and non-governmental agencies, many of which obtain at least part of their funding from national aid agencies. The two main types of aid agencies that provide public finance for infrastructure development projects in developing countries are multi-lateral and bilateral aid agencies.

Multi-lateral aid

Multi-lateral aid is a major source of official finance for infrastructure projects, and is supplied through various multi-lateral development banks (MDBs). The major development banks include the World Bank Group, the Asian Development Bank (ADB), the African Development Bank (AfDB), and the Inter-American Development Bank. These are all multi-lateral funding agencies, drawing their funds from several different countries. They operate as commercial banks, lending money at agreed rates of interest. The loans have to be repaid, but the loan conditions are often more favourable than for commercial banks, and they often allow a period of grace before repayments commence.

The actual size of disbursements of the MDBss does not reflect their overall importance. This is because MDBs act as a financial catalyst. The International Finance Corporation (IFC), for instance, frequently syndicates project financial packages from commercial banks and insurance companies, but only its own funds appear statistically as MDB disbursements. The World Bank generally finances only about one-third of the capital costs of its projects and so its influence is on three times the total amount of its disbursements. This is discussed under financial institutions in the next chapter.

The United Nations agencies

The United Nations Development Programme (UNDP) was created in 1966 as a permanent organ of the General Assembly and is funded through contributions from member countries. The UNDP provides finance and technical assistance directly and through associated organisations such as the World Bank, the International Labour Organisation (ILO) and the International Fund for Agricultural Development (IFAD).

The UNDP is the central organ of the United Nation's development aid and has two main subsidiaries: the United Nations Special Fund (UNSF) and the Expanded Programme of Technical Assistance (EPTA). The UNSF mainly provides pre-investment finance for multi-national projects, while the EPTA provides technical assistance to projects in developing countries. The United Nations Industrial Development Organisation (UNIDO) is another United Nations agency that finances and provides technical assistance to developing countries. The main purpose of UNIDO is to promote and accelerate the industrialisation of developing countries through direct assistance and mobilisation of international resources.

Bilateral aid

Bilateral aid is finance provided by a foreign government to assist in funding the recipient country's projects. Bilateral aid results in capital flow from individual donor governments in developed countries to developing countries. The objectives of bilateral donors while providing bilateral capital flow range from increasing ties and gaining political influence to ideals such as fighting poverty and environmental protection. The political and geopolitical importance of a developing country often play a big role in motivating developed countries to provide bilateral finance to developing countries.

Most industrialised countries have their own government bilateral aid agencies such as the UK's Department for International Development (DfID), the Overseas Economic Co-operation Fund (OECF) of Japan and Kreditanstalt Fur Wiederaufbau (KFW) of Germany. These agencies fund projects in developing countries through loans and grants, and also direct some of their allocated funds to those development banks of which they are members. Aid awarded directly by these agencies is often *tied aid*, i.e. the grant or loan is conditional on some of the goods and services needed for the project being procured from the donor country. This condition often means that the project is implemented by contractors and suppliers from the donor country.

By definition, tying aid precludes international competition in procurement. This varies from one country to another, but in general terms suppliers normally know when tied aid is being used and they often take advantage of the situation by introducing an element of monopoly because the buyer is not able to procure goods from elsewhere.

The aid route to raising finance can be long and time consuming, with no certainty that aid will be awarded. One of the major disadvantages of public funding provided by both multi-lateral and bilateral agencies is that there is excessive bureaucracy that adds to the cost of the project (*bureaucratic cost*). In many cases, investments are made putting more weight on the political and macro aspects of donor relations than on the micro aspects of the project. This is mainly because the repayment of the project loan is part of the public debt and is not tied to project revenues. This means that a project financed from public funds (provided by multi-lateral and bilateral donors) can be a failure, and yet the loan will still be repaid from public funds generated through general taxation.

The Department for International Development (DfID)

The DfID is part of the Foreign and Commonwealth Office and manages Britain's programme of aid to about 140 developing countries. The DfID states that its main aim is to work above all to reduce poverty worldwide and to foster sustainable development where needed, through the financing of projects, materials, equipment, technical advise, training and research. At present, Britain provides about 55% of its aid bilaterally, three-quarters of which goes to the poorest countries, mostly in the form of grants. This bilateral funding gives rise to procurement from British Exporters. The remaining 45% of aid is provided through contributions to multi-lateral agencies.

The Overseas Project Fund

The fund is administered by the Department of Trade and Industry (DTI). It provides partial grant support for activities such as feasibility studies and the preparation of tenders undertaken when pursuing large (normally more than £50 million) overseas capital projects.

The Commonwealth Development Corporation (CDC)

Established in 1948, the CDC is part of the UK's aid programme. It is similar in concept to the IFC in that it provides long-term loans and risk capital to new and existing private sector projects in overseas markets. The CDC was originally restricted to financing Commonwealth countries, but it now operates in 69 countries. Its activities are concentrated in developing countries. Its investments are normally below 35% of the project cost, and vary in size between about £0.5 million and £30 million.

The Caisse Central de Cooperation Economique (CCCE)

The CCCE, of France, is similar to the CDC in that it is a public corporation. It extends long-term project loans to developing countries. It lends at favourable rates to the poorest developing countries and at market rates to others. It requires government guarantees of repayment.

The Kreditanstalt Fur Wiederaufbau (KFW)

The KFW of Germany is a public corporation which extends project loans and technical assistance on soft loan terms to developing countries. It lends to developing countries at relatively low interest rates. It requires government guarantees of repayment.

The German Development Company (DEG)

The DEG is a non-profit-making firm which supports private investment in developing countries. It operates on commercial principles, and it extends project loans, participates in the enterprise's equity capital and offers advice to firms in developing countries.

Export–Import Bank (EXIM Bank)

The EXIM Bank of Japan extends loans where private lenders refuse to invest, at market terms and rates of interest.

The Overseas Economic Co-operation Fund (OECF)

The OECF of Japan finances projects in developing countries, with varying rates of interest depending on the borrower's situation and the project type. The OECF finances projects beyond the scope of the EXIM Bank.

The Japan International Co-operation Agency (JICA)

The JICA extends grant aid, soft project loans, and technical and training assistance at favourable rates in developing countries. The JICA extends assistance even in areas where the EXIM Bank and the OECF find it difficult to invest because of high risk, low return or technical obstacles.

The Netherlands Investment Bank for Developing Countries

This bank extends grants and project loans to developing countries. Its grants are tied to home country goods and services, while its loans are less concessional. The rates and terms of its lending vary with the level of development of the recipient country.

The Netherlands Finance Company for Developing Countries (FMO)

The FMO is another Dutch source of project finance and aid for developing countries. It provides mainly technical assistance, equity investment and project loans to the private sector in developing countries. It also extends grants. Its loans are usually at market rates of interest, but procurement is not tied to home country goods and services.

The Swedish International Development Agency (SIDA)

The SIDA extends mainly grants and interest-free soft loans to developing countries. Some of SIDA's grants are tied to the procurement of home country goods and services.

The Kuwait Fund for Arab Economic Development (KFAED)

The KFAED extends project loans and pre-investment grants to developing countries around the world. It requires government guarantees of repayments, and its conditions of lending vary from soft to hard.

The Abu Dhabi Fund for Arab Economic Development (ADFAED)

The ADFAED of the United Arab Emirates extends project loans, equity investment and technical assistance to developing countries. Its loans are on medium terms and at relatively favourable rates of interest.

The Saudi Fund for Development (SFD)

The SFD extends project loans mainly to Arab and Islamic countries. It lends to various sectors, with infrastructure receiving most of its funds.

Co-financing

Co-financing is an important way of enhancing the resources of multi-lateral development banks. There are three main types of co-financing partners: other MDBs or funds, particularly those from oil rich countries; export credit agencies, which are directly associated with the finance of goods and services from a particular country; and commercial banks. Co-financing will usually be undertaken when the project meets the objectives of the borrower, the co-financiers and the principal MDBs.

Alternative methods of financing infrastructure projects

To date, a lot has been done to develop infrastructure in developing countries, but infrastructure development is neither keeping up with the demand (quantity and quality) nor is it sustainable in most cases. More infrastructure is required in terms of quantity and quality.

In recent decades, developing countries have made substantial investments in infrastructure, achieving dramatic gains for households and producers by expanding their access to services such as safe water, sanitation, electric power, telecommunications and transport. Even more infrastructure investment and expansion are needed in order to extend the reach of services to those who are not yet provided for. It is important to consider the quality of infrastructure, even as developing countries concentrate on quantity. Low operating efficiency, inadequate maintenance and a lack of attention to the needs of users have all played a part in reducing the development impact of infrastructure investments in the past. Infrastructure development projects that are implemented but not operated and managed efficiently are not sustainable. Both quantity and quality improvements are essential to modernise and diversify production, help countries compete internationally, and accommodate the rapid urbanisation prevalent in most developing countries.

There exists a direct link between infrastructure and development. There is a need to explore ways in which developing countries can improve both the provision and the quality of infrastructure services. Infrastructure is an area in which government policy and finance have an important role to play because of its pervasive impact on economic development and human welfare. While the special technical and economic characteristics of infrastructure give government an essential role in its provision, dominant and pervasive intervention by governments has in many cases failed to promote efficient or responsive delivery of services. Where infrastructure services are financed and managed exclusively by the public sector, these services are the first to suffer when governments are faced with financial problems.

Traditional methods of financing infrastructure projects have proved both inadequate and unsustainable. Public financing through general taxation is inadequate because developing countries are poor. Their economies are unable to support the development of infrastructure required by their population. Public financing of infrastructure projects through finance provided by international agencies has reduced throughout the 1990s. In addition, public financing of all infrastructure projects is not sustainable, as it results in a great debt burden to developing countries.

Most developing countries are now experiencing this *reverse transfer* of resources, since they are spending a high proportion of their foreign exchange in repayment of debt. The debt burden follows the receipt of relatively large amounts of loans since the 1980s. Official development assistance in the form of loans and grants from the traditional sources has reduced greatly at a time when financial resources are needed most. It is, therefore, necessary for developing countries to seek private financing and management of infrastructure projects.

There is need to look for a third method of financing by involving the private sector, especially for projects that have potential for a revenue stream. It may be stated that if growth rates in developing countries are to increase and be maintained at a suitable level, private sector funding must play a larger role.

In contrast to the centrally planned approach to project identification and finance in the public sector, most private sector projects originate through the perception of a market opportunity. Governments worldwide should provide incentives to private investors and realise that such incentives do not necessarily compromise the attainment of other objectives. Privatisation is an alternative method of financing and managing infrastructure projects, and this is discussed in Chapter 3.

Reducing country risks to improve the environment for private sector development

Prosperity is usually expressed in terms of sustainable economic growth and poverty reduction. Evidence from research has shown that growth and development in particular are functions of a good macro-economic policy framework, which facilitates capital flows to productive economic activities. An important point is to understand the factors that determine capital flows to different countries and regions, and what impact they have on their development process.

The largest global source of investment capital is the private sector. *Foreign direct investment* (FDI) plays a significant role in the development process of most developing countries. FDI is at present more important than official development assistance (official development assistance) or foreign aid.

During the past decade, FDI to developing countries has increased considerably, from 0.8% of GDP in 1991 to 2.0% in 1997. Furthermore, developing countries increased their share of global FDI flows from 21% to 36% during this period. This contrasts with the static or declining flow of FDI for most African countries, together with declining trend of development aid, especially to Africa, which is the poorest continent.

Aid per capita to many developing countries has dropped sharply in the last decade, essentially because rich nations commit only 0.24% a year of their GDP to aid, although they have a long-term commitment to provide 0.7% of GDP. A new World Bank (1998) report, *Aid and Reform in Africa – Lessons From Ten Case Studies*, underlines the case for increasing foreign aid to poor African countries, and particularly to those that are pursuing sound economic policies. Since foreign aid is a small proportion of international capital flows, it is essentially supposed to assist a country to create a conducive environment for attracting private investment.

FDI is important, not just for the financial flows to countries which otherwise do not have large pools of domestic capital, but also because it facilitates the transfer of technology and managerial know-how. However, foreign investors are highly selective and will only invest in areas where they can forecast good returns. This is clearly illustrated by the concentration of FDI flows in developing countries that have liberalised their economies and provided a good environment for the private sector – China, Brazil, Mexico, Chile and India are a few examples. Africa, unfortunately, receives a very small percentage of international investment flows. We may then ask ourselves: What attracts foreign investors?

Factors that attract foreign direct investment are:

- a liberalised economy, with minimal state involvement in commercial enterprises
- a strong GDP and incomes growth, leading to market potential for investors' goods and services
- economic, social and political stability
- a predictable policy framework
- efficient infrastructure (roads, telecommunications, power, railways, etc.)
- efficient public sector administration, including the legal sector, and low levels of corruption.

These are not the only factors, but they are the major ones. These factors translate into risks that investors have to deal with when considering investment. Countries that create a favourable environment for private sector growth benefit most from FDI. A number of developing countries are providing such conditions, and with the implementation of economic reforms the trend will hopefully change in the future. In some countries, private investment is low because economic, social and political uncertainty has made it difficult to build private sector confidence. Reforms deal with difficulties, seen as risks, that constrain the viability of business ventures. They can greatly improve the investment climate, and therefore enable countries to attract considerable private capital for investment in productive economic activities.

Surveys of potential investors in developing countries show that they want the following:

- improved economic governance and increased efficiency in the judiciary and civil service
- budget reforms that will ensure maximum utilisation of tax revenue and other funds available to the public sector (e.g. official development assistance)
- privatisation of key public utilities, especially in energy, telecommunications, railways and water supply – not just opening up opportunities for new private capital investment but, equally importantly, reducing the cost of doing business through better and more reliable services
- extensive consultations between the government and other stakeholders on the development process, based on a home-grown development strategy
- increased efficiency of the financial sector in order for commercial banks to reduce their bad debt overhang and thus reduce domestic interest rates
- restrained commercial borrowing by the government (to keep interest rates down)
- dealing with insecurity, which not only affects the cost of doing business, but also can be a real deterrent to foreign investor interest.

Developing countries need to address some or all of the above issues in order to enhance their standing in the global competition for foreign investment. In many developing countries there are plenty of prospects and strong points to encourage investors. It should be noted, however, that investors typically invest in areas where the chances of obtaining competitive returns on investment are high.

Bibliography

International Finance Corporation. *Emerging Stock Markets Factbook, 1995.* The International Finance Corporation, Washington, DC, 1995.

Merna, T. and Njiru, C. *Financing and Managing Infrastructure Projects.* Asia Law and Practice, Hong Kong, 1998.

United Nations Development Programme. *Human Development Report.* Oxford University Press, New York, 2000.

World Bank. *World Development Report, 1994: Infrastructure For Development.* Oxford University Press, Oxford, 1994.

World Bank. *Aid and Reform in Africa – Lessons From Ten Case Studies.* Report. Worldbank.org, 1998.

Wells, S. J. *International Economics.* Minerva, 1973.

CHAPTER TEN

Financial institutions

Introduction

This chapter describes the different sources of finance used in financing projects. Different financial institutions have different lending terms and provide different types of capital. The chapter identifies a number of the major financial institutions involved in financing projects and outlines the types of financial products available.

Financial institutions

Commercial enterprises can be classified as either financial or non-financial. *Non-financial enterprises* are either manufacturing (e.g. textiles, white goods) or service providers (e.g. transportation, telecommunication). *Financial enterprises* provide one or more of the following functions:

- financial intermediation
- brokerage
- managing funds
- underwriting securities
- guarantees.

Financial intermediation has remained the most important function of financial institutions. It refers to acquisition of financial assets from the market and transforming them into more preferable types of assets for use by the ultimate investors. For example, commercial banks accept deposits from the general public and thereafter lend the money to projects or enterprises who are in need of funds. Similarly, a mutual fund or unit trust receives funds from the public and invests in projects to generate sufficient resources to pay to its lenders and also make a surplus. The financial intermediation function is performed by depository institutions and non-depository institutions. *Depository institutions* are institutions that receive funds from the public and companies in the form of deposits. Depository institutions include retail banks, wholesale banks, savings banks and building societies. *Non-depository institutions* include insurance companies, pension funds and investment institutions (unit trusts, investment trusts, property based funds, other managed funds). Non-depository financial institutions are primarily concerned with the management of the funds of their clients, but in the process they also perform the intermediation function.

When a new issue of shares or bonds is issued, investment bankers underwrite the issue. This means that if the securities are not sold fully the remaining securities are placed with the underwriter. The underwriter charges a commission for this service. In large lending by banks and other financial institutions, particularly to developing countries where there are higher risks involved, the lenders insist on a guarantee for repayment. Such guarantees may be provided by: a bank; an agency specially set up for this purpose, such as an export credit guarantee organisation; a multi-lateral financial institution, such as the World Bank; or the government of the country where the borrowing entity is located. Brokers and dealers primarily operate in the secondary market.

From a project financing point of view, we may classify financial institutions into two broad categories: domestic and international.

Domestic financial institutions

Domestic financial institutions can be considered to comprise the following broad categories:

Depository institutions (commercial banks, savings banks, building societies)

Both commercial banks and savings banks collect deposits from the public and make direct loans to various entities. They also invest in shares. Building societies also collect deposits from the public that are mostly loaned as property mortgages. Only the surplus funds are loaned for other purposes.

Non-depository institutions (pension funds, insurance companies, unit trusts)

The funds received by pension funds and insurance companies are long term in nature. They are therefore in a position to lend for long periods or invest in equity. Unit trusts are primarily involved in the management of funds, and buy and sell equities in the equity market to make a profit for the purchasers of the units of the unit trust.

Investment or merchant banks

These specialise in:

> ... advising on, arranging, placing and underwriting new capital issues, acting as intermediary between companies involved in mergers and acquisitions and acting as consultants with expertise in investments and portfolio management. (Piesse, Peasnell and Ward, 1995)

In the UK these banks were traditionally involved in guaranteeing the payment of trade bills, when delivery dates and capital transfers were much more complicated than they are now. They continue to have this function even now.

Infrastructure development banks

Banks of this type have been set up in many countries, in particular to meet the requirements of infrastructure financing. They specialise in projects such as water, solid waste collection and disposal, and roads. For example, in Japan the Japan Development Bank (JDB) has been crucial in the development of infrastructure in the past, and even now it continues to play a major role. It has relied largely on postal savings to fund its operations. In Europe, municipal banks that obtain their resources from contractual savings and other long-term sources have been performing the role of infrastructure financiers.

In developing countries, however, infrastructure development banks have generally suffered from the public ownership of these banks. This has led to inefficient targeting, subsidisation of lending, interference in operations and corruption. They have also continued to rely on government funding for their operations. However, the *World Development Report, 1994* (World Bank, 1994) reports that some specialised infrastructure intermediaries are performing well and are also assisting in the development of capital markets. For example, in India, Infrastructure Leasing and Financial Services and the Housing Development Finance Corporation are aiming to sell their loans to other private financial institutions once projects have been successfully completed. Another specialised infrastructure bank, BANOBRAS in Mexico, is promoting private water and sewerage projects by guaranteeing that municipalities will pay for the services provided.

In certain countries infrastructure funds are being set up, as a transitional mechanism, to support the development of infrastructure until long-term financing mechanisms such as long-term development banks and capital markets are developed. These funds are either government sponsored or private. For example, the Private Sector Energy Development Fund in Pakistan and the Private Sector Energy Fund in Jamaica are designed to catalyse private financing for power projects. In Thailand, a Thai Guarantee Facility for financing environmental infrastructure has been set up. It will not provide any financing, but will guarantee private loans to municipalities and other private operators. The Regional Development Account in Indonesia is a transitional credit system.

Venture capital funds

Venture capital funds are used for start-up ventures, expansion and growth. There are two formal sources of venture capital in the UK. The Investors in Industry plc, otherwise known as the 3i Group, was set up by a group of institutions including the Bank of England. The 3i Group provides a number of specialist financial services, including the provision of capital to firms expecting to expand quickly. It provides financing either in the form of a loan or by buying shares. The other sources include independent firms exclusively concerned with the provision of venture capital, such as Northern Venture Managers and Venture Capital Subsidiaries which were set up by the clearing banks.

International financial institutions

The globalisation of financial markets has led to enormous growth in overseas banking and an increase in intermediation between the financial institutions of one country and the suppliers of investment funds and the user of these funds from other countries. The growth of the offshore market, or Euro-currency market, in recent years has been phenomenal. The role of multi-lateral and bilateral financial institutions has also increased significantly. Nevitt (1983) lists the following advantages and disadvantages of international financial institutions:

Advantages:

- loans are for longer term than other sources
- loans carry lower interest rates
- loans act as an endorsement for credit from other sources
- loans act as a basis for co-financing arrangements.

Disadvantages:

- there is a lengthy approval process
- the loans are in hard currency, which provides little opportunity for optimising currency risk.

The following international financial institutions are important when considering the financing of infrastructure projects.

The World Bank Group

The World Bank Group plays an important and visible role in supporting project finance operations all over the world, with special emphasis on developing countries. It consists of the following five separate, but affiliated, organisations:

- the International Bank for Reconstruction and Development (IBRD)
- the International Development Association (IDA)
- the International Finance Corporation (IFC)
- the Multi-lateral Investment Guarantee Agency (MIGA)
- the International Centre for the Settlement of Investment Disputes (ICSID).

The World Bank is the largest multi-lateral agency that provides public finance for public projects in developing countries.

The World Bank itself is made up of the International Bank for Reconstruction and Development (IBRD) and its affiliate the International Development Association (IDA).

The IBRD, along with the International Monetary Fund (IMF), was born at an international conference held in Bretton Woods, New Hampshire in 1944. The World Bank was set up in 1945–1946 and made its first loan in 1947. At first, the World Bank made loans to European nations for post-War reconstruction, but soon (1948) began lending to southern hemisphere countries as well, for development.

The World Bank's long-term objective is stated as the establishment of efficient private enterprise systems in developing countries. Its primary function was to help finance essential infrastructure that could not be financed by private markets, but its broader development strategy was to persuade governments to establish sound policies of macro-economics management and neutral markets, free of government intervention and price distortion, that would function efficiently and equitably. The thinking was that, once these conditions were in place, free markets would flourish and private enterprise would develop without the need for direct bank or government intervention.

The World Bank has over the years assumed a pre-eminent global role and reputation in development finance and virtually unchallenged leadership in the design and implementation of development programmes, strategies and policies in the developing world. This has been possible because the Bank has strengths in many areas such as:

- It has a large membership and a large capital base.
- It has an international image.
- It has financial influence over the regional and bilateral development agencies. Countries with good relations with the World Bank are able to benefit from additional resources co-ordinated by the World Bank. The Bank plays an important role in sourcing the finance for a project even when it cannot itself finance the full amount.
- It has a conservative lending policy. The World Bank lends only to sovereign states. The IBRD loans typically include a 5-year grace period, after which borrowing countries have 15–20 years to pay back at market interest rates, which today average about 8%. The IDA credits are virtually interest free (about 1% service charge) with a 10-year grace period and a repayment period of about 50 years. The terms are indeed soft, but the loan principal must be repaid as the credits are not gifts. The IDA's funds are replenished from time to time by donor countries as part of their overseas development aid. The World Bank has enormous influence over other public and private sources of capital, and so its clients put the Bank at the top of their lists of creditors to be repaid.
- It has a top credit rating that enables it to raise money from the international money markets and lend at relatively low interest to borrowing member countries. The Bank does not finance private sector projects that are not guaranteed. It has a rigid sovereign guarantee requirement when financing private sector projects through its affiliate the IFC.
- It provides untied aid.
- It provides finance for infrastructure development.
- It has international power and stature.

The political preoccupation and concerns of the rich industrialised countries, which are its major shareholders, often influence the World Bank's agenda. The Bank applies its agenda by imposing conditionalities on developing countries before lending the much needed funds for development projects.

The International Bank for Reconstruction and Development

The IBRD was established in 1944 to promote the economic development of developing countries through the provision of financial resources and technical advice. It is the single largest provider of development loans to middle-income developing countries. The IBRD approved new lending commitments of US $19.1 billion for the fiscal year 1997. The IBRD participates in project finance operations through a variety of instruments: debt, equity, guarantees, political risk insurance and co-financing.

Debt

Debt financing is the predominant instrument employed by the IBRD to support project finance. Loans are structured to cover a specified period of time, which covers the construction phase and some initial years of operation. The loan is either provided directly to the project company, in which case it receives a guarantee from the relevant country, or through the government which lends to the project. Sometimes the IBRD provides loans to countries, which are eligible only for soft term IDA credits, if there is a specific promising project expected to produce substantial foreign exchange to repay the loans. Such projects are known as *enclave projects*. Lending to an enclave project could be either directly to the project company, in which case the IBRD receives a counter guarantee from the relevant country, or through the government which lends to the project.

Equity

The IBRD does not make equity investment directly to a project company. However, it can make a loan to a country to be used by that country to finance its equity in the project company.

Guarantees

The IBRD, in recent years, has expanded its support for guarantees for investors. It provides guarantees in two distinct ways:

* through the issue of a guarantee directly to lenders to a project
* by financing through loans the guarantee coverage issuance by a country.

The former is a direct form of guarantee and the latter an indirect financing of guarantee. Recently, the IBRD has adopted a programme to expand its guarantees, in particular to cover infrastructure projects, known as *mainstreaming of guarantee*. Under this facility it provides two types of coverage:

* a partial risk guarantee protecting the repayment of loans against the non-performance of sovereign contractual obligations or from *force majeure*
* a partial credit guarantee covering a portion of the payments due under loans.

The first type of guarantee is more suitable for limited recourse financing such as build–own–operate–transfer (BOOT) projects.

The partial credit guarantee can also be implemented through a put option to the holder of long-term project debt, which will give him the right to sell the loan to the IBRD after a number of years.

Political risk insurance
The IBRD can make loans to a country, to be used to finance political risk insurance, covering either debt or equity.

Co-financing
Co-financing refers to any arrangement under which funds from the World Bank are associated with funds provided by other sources outside the borrowing country in financing a particular project. Official sources of co-financing include member governments, their agencies and multi-lateral financing institutions. There can be co-financing by private institutions, such as banks, insurance companies and other capital markets outside the country of the borrower.

The International Development Association
The IDA was established in 1960 to make soft loans to the world's poorest countries that cannot afford the IBRD terms. The IDA formally calls its loans *credits* to distinguish them from the IBRD's loans. The credits lent by IDA have soft repayment terms (virtually interest free, a service charge of about 1%, a 10-year grace period and a repayment period of about 50 years), but the loan principal must be repaid. The IDA's funds are replenished from time to time by donor countries as part of their overseas development aid. Many countries receive both Bank loans and IDA credits.

The IDA's purpose is to promote the economic development of these countries. It approved over US $4.6 billion in loans during its 1997 fiscal year. The IDA participates in projects through development credit agreements. The IDA credits for a project are always made through the government, never directly to a project company. The government, however, lends it to the project company. Just like the IBRD, the IDA also does not make equity investment directly to a project company. However, it can make loans to a country to be used by that country to finance its equity in the project company. The IDA has the potential power to give a guarantee, but it has not exercised this so far. The IDA has not been authorised to finance loans for risk insurance coverage.

There is a noticeable reduction in the level of finance provided by the World Bank to most developing countries. The Bank has recommended that governments should downsize as a way of reducing public expenditure and containing the large debt burden. There is now emphasis for developing countries to involve the private sector in the provision of infrastructure finance and management. The focus has increasingly moved to the World Bank's sister organisation, the IFC, which deals more with the private sector.

The International Finance Corporation

The IFC was established in 1956 to complement the support provided by the IBRD. Whereas the IBRD and the IDA operate through governments, even when supporting private sector operations the IFC provides direct support to the private sector, without government financial guarantees.

The IFC is able to encourage and support development in the private sector by financing worthwhile new projects and also by rehabilitating potentially good, financially distressed enterprises. The IFC's objectives, capabilities and catalytic role can help fill the gap left by the export credit agencies, banks and private sources of direct equity investment. The ability of the IFC to invest in equity and to provide project loan finance to ventures without seeking financial guarantees from the host government complements the role of the multilateral development banks (MDBs). The evolution of a concept into a defined project and the development of the institutional capacity to manage the project requires a strong and wilful promoting group.

The criteria by which the IFC selects new projects are based on judgement and market principles. Projects in developed countries cannot be considered, although projects may be introduced by an entrepreneur in a host country to potential foreign partners and he may choose any stage in the project cycle to talk to the IFC about it. The IFC interest in an investment opportunity then acts as a catalyst for other institutional and private investors who would otherwise have little interest in the project as they would perceive the project as highly risky. The IFC is therefore an investment bank, but with a development focus.

Initially, sponsors often prefer to introduce a project to the IFC informally, but written information at an early stage is required. This information should provide the basic parameters of the project and the rationale behind it, as well as the identity of the proposing group. If the project seems viable after a formal initial evaluation by the IFC, it is developed in phases dictated by commercial pressures, unlike the deliberate pace of development of public sector projects. After passing the initial evaluation, the project usually passes through a further phase of restructuring by the project sponsor aided by the IFC, after which the project is ready for appraisal.

As IFC investments are wholly dependent on the financial success of a project coupled with the absence of government guarantees, the appraisal is more commercially orientated than a similar public sector project. The financial plan is studied in great detail to ensure that it is adequate to complete the project with gearing that provides safe debt service coverage at an acceptable ratio. The IFC appraisal report assesses the project risks, identifies issues that must be resolved before the investment is made and states the financial and economic rates of return that should be expected. The appraisal report is given to the board, who uses it to make a decision on whether or not to invest in the project.

Most developing countries have not been using the IFC facility for development of infrastructure by the private sector, possibly because there has been ready public finance from the World Bank. The situation is expected to change with the reduction of public finance from multi-lateral and bilateral donors. Developing countries will need to actively involve the private sector in the financing and management of infrastructure projects if the challenge of infrastructure provision is to be met.

In the fiscal year 1997, the IFC approved 276 projects, investing US $3.3 billion. Unlike the IBRD, the IFC is not limited to making loans or issuing guarantees of loans. It can make investment in any form it deems fit, including equity, loans, convertibles and guarantees.

Loans

Loans by the IFC are always provided directly to a private sector entity. The terms of the loan are essentially commercial. In addition to loans from its own resources the IFC also syndicates loans (referred to as *B-loans*). These are funded by commercial lenders under the umbrella protection of the IFC.

Equity

The IFC provides substantial equity investment in enterprises. However, it generally assumes a passive role. It is prohibited by its articles to assume the responsibility of management.

Quasi equity

The IFC also makes investments in instruments such as preferred stock and convertible debentures.

Guarantees

The IFC also provides guarantees to lenders. However, this activity is a small proportion of its investment portfolio.

The Multi-lateral Investment Guarantee Agency

The MIGA was established in 1988 to encourage the flow of foreign investment to developing countries, mainly through the provision of political risk insurance. Many countries have their own national insurance programmes, but this generally covers investment by its own nationals. The MIGA has augmented the availability of political risk insurance, while avoiding country-specific restrictions. During 1995 the MIGA issued political risk coverage totalling $670 million. The MIGA is empowered to issue guarantees, including co-insurance and re-insurance, against non-commercial risks, for investments. The MIGA can provide a guarantee against losses resulting from currency convertibility, expropriation, war and civil disturbance, and also breach of contract.

The International Centre for the Settlement of Investment Disputes
The ICSID provides facilities for the conciliation and arbitration of disputes between governments (or constituent subdivisions or public agencies) of ICSID member countries and investors (individuals or companies) that qualify as nationals of other member countries. The Centre's objective in making such facilities available is to promote an atmosphere of mutual confidence between governments and foreign investors that is conducive to private international investment.

The African Development Bank

The African Development Bank Group is a multi-national development bank supported by 77 nations (member countries) from Africa, North and South America, Europe and Asia. The Bank Group consists of three institutions: the African Development Bank (AfDB), the African Development Fund (ADF) and the Nigeria Trust Fund (NTF).

The African Development Bank

Established in 1964, the AfDB has a mission to promote economic and social development through loans, equity investments and technical assistance of its regional member countries. The Bank's authorised capital is US $23.29 billion. The financial resources of the Bank consist of subscribed capital, reserves, funds raised through borrowing from international money and capital markets, and accumulated net income. The Bank's operations cover a large number of sectors, with particular emphasis on agriculture, public utilities, transport, industry, health and education, poverty reduction, environmental management, gender mainstreaming and population activities. Most Bank financing is designed to support specific projects. However, the Bank also provides programme, sector and policy-based loans to enhance national economic management. The Bank also finances non-publicly guaranteed private sector operations. The Bank actively pursues co-financing activities with bilateral and multilateral institutions.

The Bank lends at a variable lending rate calculated on the basis of the cost of borrowing. The rate is adjusted twice a year, on 1 January and 1 July, to reflect changes in the average cost of borrowing over the preceding 6-month period. For 1996, the rate was 7.50% for January–June and 7.31% for July–December. The other terms include a commitment charge of 1.00% and maturities of up to 20 years, including a 5-year grace period.

The African Development Fund

The ADF provides development finance on concessional terms to low-income regional member countries (RMCs) that are unable to borrow on the non-concessional terms of the AfDB. In accordance with its lending policy, poverty reduction is the main aim of the ADF's development activities in the borrowing countries.

The ADF finances projects and technical assistance, as well as studies. It lends at zero interest rates, with a service charge of 0.75% per annum, a commitment fee of 0.50%, and a 50-year repayment period, including a 10-year grace period.

The Asian Development Bank

The Asian Development Bank (ADB) was established in 1966, and has 56 member countries. It is a development finance institution, which primarily lends funds and provides technical assistance to developing member countries. It is engaged in promoting the economic and social progress of its developing member countries in the Asian and Pacific region, and is owned by the governments of the 40 countries from the region and 16 countries from outside the region. The principal functions of the ADB are:

- to make loans and equity investments for the economic and social advancement of developing member countries
- to provide technical assistance, including advisory services, for the preparation and execution of development projects and programmes
- to promote investment of public and private capital for development purposes

- to respond to requests for assistance in co-ordinating development policies and plans of developing member countries.

The ADB financing includes loans, equity investments, and guarantees. Since the USA is a shareholder and contributor to the ADB, US companies are eligible to take part in ADB funded projects.

Requirements

ADB supported private sector projects must be economically viable, create jobs and positively affect the borrowing country's economy. Preference is given to projects that assist in removing economic bottlenecks and those which transfer technology.

Eligible enterprises should be in the private sector of an ADB developing member country and may be locally or foreign owned. An enterprise owned jointly by private interests and the government may also be eligible, provided that it satisfies the criteria of operational autonomy and managerial freedom and is run on a commercial basis.

The ADB normally requires sovereign guarantees. However, the ADB makes direct loans to private companies without sovereign guarantees if the project produces essential items or provides vital services.

As a general rule the ADB does not provide majority funds for private sector projects, and avoids responsibilities associated with ownership except if this becomes necessary to safeguard the Bank's investment. The ADB tries to sell its equity in a company at a fair price as soon as possible, but consults with major investment partners and gives due consideration to their views. The total amount of ADB assistance to a project does not normally exceed 25% of the total cost of the project, or US $50 million, whichever is lower.

Facilities

Direct financial assistance to private enterprises consists mainly of loans without a government guarantee, or underwriting and investment in equity securities. Direct assistance is also provided to privately owned financial institutions. The ADB has underwritten the initial offerings of several mutual funds. Financial institutions and sponsors of projects involving venture capital, leasing, factoring, investment management and commercial finance, among others, are eligible for the Bank's direct assistance. In addition to assisting financial intermediaries, the ADB's private sector operations focus on infrastructure projects and industrial, agribusiness and other projects that have significant demonstrational or economic merit.

Loans

Loans without government guarantees have been made available by the ADB since 1985. Direct financing is largely allocated to capital-intensive infrastructure projects. Credit lines established with financial institutions, by the ADB, for on-lending are intended mainly for small and medium-sized new ventures as well as for balancing, modernisation, replacement or expansion of existing ventures. The Bank has several different loan products, with maturity generally ranging from 8 to 15 years, including a suitable grace period.

Market based lending

Market based lending (MBL) loans are more appropriate for borrowers that can pass on the foreign exchange risk to others, such as financial intermediaries. MBL loans allow both financial intermediaries and their sub-borrowers to match the currencies of their cash inflows and outflows. In addition, the market-based lending rate may suit the needs of financial intermediaries that usually use the London Interbank Offer Rate (LIBOR) as the base rate for their foreign currency lending.

US dollar loans

US dollar loans are appropriate for borrowers that earn revenues in US dollars or in currencies that tend to move with US dollar rates and do not require close linkage to current market interest rates. The variable interest rate of US dollar loans is based on the cost of the Bank's outstanding US dollar borrowing and is relatively stable.

Multi-currency loans

Multi-currency loans present borrowers with a significant exchange risk and are difficult to manage because of the nature of the currency composition. Consequently, these loans are appropriate only for borrowers that have the capacity to manage such currency risks. The variable lending rate of these loans is not linked to current market interest rates, but rather is based on the average cost of the Bank's outstanding borrowing undertaken to fund these loans.

Equity investments

ADB equity investment operations are intended to complement domestic resources and co-financing. The Bank's equity operations takes the form of investment in the equity of productive enterprises for financing specific projects, or investment in the equity of development finance institutions and other institutions set up to promote industry, mining, agribusiness or allied activities.

Guarantees

The ADB's guarantees offer limited support designed to reduce private sector exposure to risk, unique to developing countries, that the private sector cannot absorb or manage on its own. Guarantees are most appropriate in supporting private funding of infrastructure projects. These projects require considerable funds with extended maturity to match their long pay-back periods. The Bank's guarantees can be offered in any currency for commercial debt financing to private entities in ADB developing member countries. The types of guarantee in operation are the partial-credit guarantee and the partial-risk guarantee.

Partial-credit guarantees are intended for private sector borrowers with a government counter-guarantee. These are designed to cover the portion of financing that falls due beyond the normal tenure of loans provided by private financiers. This mode is generally used for private sector projects that need long-term funds to be financially viable. A partial credit guarantee typically extends maturities of loans and covers all events of non-payment for a designated part of the debt service. In that sense, it is an all-inclusive guarantee, either of principal or principal and base interest, for those maturities that cannot be obtained from commercial lenders without credit enhancement. This facility is particularly useful in ADB developing member countries with partially restricted access to international capital markets, but which are considered fundamentally creditworthy and sound by the ADB, for obtaining longer maturities of credit needed to improve project viability.

Partial-risk guarantees are provided to cover specific risks arising from non-performance of government contractual obligations that are critical to the viability of projects. Partial risk guarantees can mitigate specific risks that private financiers generally find difficult to absorb or manage. Such guarantees typically cover risks arising from government actions, such as, non-delivery of inputs and/or non-payment for output by state-owned entities, changes in the agreed upon regulatory framework, and political *force majeure*. Foreign exchange convertibility risk in projects that do not generate foreign exchange earnings may also be covered. A government counter-guarantee is insisted on to reaffirm the government's acceptance of the obligations backed by the ADB.

The Public–Private Infrastructure Advisory Facility

The Public–Private Infrastructure Advisory Facility (PPIAF) is a multi-donor technical assistance facility aimed at helping developing country governments improve the quality of infrastructure through private sector involvement. It was developed at the joint initiative of the governments of the UK and Japan, working closely with the World Bank Group. The PPIAF was set up in 1999 and is managed by the World Bank on behalf of participating donors. The PPIAF's mission is to help eliminate poverty and achieve sustainable development in developing countries by facilitating private sector involvement in infrastructure.

The PPIAF pursues its objectives through two mechanisms:

- channelling technical assistance to governments in developing countries on strategies and measures to tap the full potential of private sector involvement in infrastructure
- identifying, disseminating and promoting best practices on matters related to private sector involvement in infrastructure in developing countries.

The PPIAF can support private sector involvement in the financing, ownership, operation, rehabilitation, maintenance and/or management of eligible infrastructure services that include the following:

- electricity generation, transmission and distribution
- natural gas transmission and distribution
- water and sewerage services provided by or through utilities
- solid waste at metropolitan level (over 500 000 residents)
- telecommunications
- railways
- ports
- airports
- roads.

PPIAF support covers a broad array of contracting approaches, ranging from management contracts and leases through concessions and divestitures. Among other factors, PPIAF support is meant to result in a net additional flow of resources to relevant activities. The PPIAF can provide funding when the proposed activity cannot obtain funding more conveniently from other funding sources, including loans from international financial institutions, grants or a government's own resources.

The Development Bank of Southern Africa

The Development Bank of Southern Africa (DBSA) is a regional development financial institution with a mandate to provide infrastructure finance. The shareholder is the South African Government. The Bank's mission is to contribute to development by providing finance and expertise to improve the quality of life of the people of southern Africa. The Bank's focus is investment in infrastructure. The Bank lends to both public and private sector infrastructure projects. Based in South Africa, the Bank is the main regional finance institution serving the Southern Africa Development Co-operation (SADC) countries. The main financial products are debt, equity and credit enhancement (guarantees).

The Commonwealth Development Corporation (CDC)

The CDC was founded in 1948 as the Colonial Development Corporation. In 1963 it changed its name to the Commonwealth Development Corporation and was able to operate in independent Commonwealth countries. In 1969 its area of operations was extended to all countries, but the Commonwealth remains the most important area. The CDC operates through the following three core business units:

- CDC Investments, which is responsible for managing the CDC's existing portfolio and for developing new business.
- CDC Industries, which operates as a management company for those businesses in which the CDC has a majority shareholding or has effective control.
- CDC Financial Markets, which carries out business development, deal structuring, due diligence, negotiation, monitoring and, where appropriate, management of the CDC's investments in the financial sector. It assists the development of small and medium-sized businesses. The CDC works with the Commonwealth Secretariat to set up special funds to invest in private sector businesses in developing Commonwealth countries.

The European Economic Community agencies

European Economic Community sources of development finance include the European Investment Bank (EIB), the European Development Fund (EDF) and the European Bank for Reconstruction and Development (EBRD). Although these agencies are multilateral in form, they operate and function as bilateral agencies, mainly serving the interests of their European Economic Community members.

The European Development Fund

The EDF was created in June 1994. The fund provides grants and soft loans to several developing countries. The EDF's funds mainly go to technical assistance programmes and pre-investment project finance for projects of various types. The Fund is an independent international financial institution, bringing together on a European scale shareholders forming a partnership between the public and the private sectors. It is a joint venture between the EIB, the European Commission and a group of banks and financial institutions from all the member states of the European Union. The creation of the Fund involved a *de facto* amendment to the Union Treaty that had to be ratified.

The mission of the Fund is to provide guarantees, essentially to the private or mixed public–private sectors, in support of long-term finance for major infrastructure projects in transport, telecommunications and energy, furthering the development of trans-European networks as well as small and medium-sized enterprises. Since 1996, the Fund has been authorised to extend its operations to equity participation.

In addition to support for small and medium-sized enterprises, the other operations of the EDF are major railway lines, including high speed trains, major urban transport systems, gas pipelines, optical fibre networks, toll motorways, bridges and tunnels, harbours, airports, electricity interconnections, data networks and mobile telephone networks.

Out of the Fund's authorised capital of ECU 2 billion the subscribed funds are ECU 1.789 billion subscribed by the European Investment Bank (ECU 800 million), the European Commission (ECU 600 million) and financial institutions (ECU 389 million).

Since the Fund was inaugurated it has approved operations amounting to ECU 2.55 billion and signed guarantees for over ECU 1.8 billion.

The European Investment Bank

The EIB was created in 1958 to promote the economic integration of the European Community. The EIB mainly finances projects in underdeveloped regions of the European Economic Community that are unable to attract conventional investment. The EIB makes or guarantees loans for financing investment principally in industry, energy and infrastructure. The EIB also provides financial assistance to countries with close links to the European Economic Community, such as Turkey and several African countries. The EIB's loans are on harder terms than those of EDF.

The EIB is the European Union's main financing institution, and provides loans for capital investment promoting the Union's balanced economic development and integration. The EIB is an enormously flexible and cost-effective source of finance whose

ECU 20 billion volume of annual lending makes it the largest international financing institution in the world.

In the European Union, EIB loans are lent to projects that fulfil one or more of the following objectives:

- strengthening economic progress in the less favoured regions
- improving trans-European networks in transport and telecommunications, energy transfer schemes for enhancing industry's international competitiveness and its integration at the European level
- supporting small and medium-sized enterprises
- protecting the environment and quality of life
- promoting urban development, safeguarding the European Union's architectural heritage and achieving secure energy supplies.

The EIB's financing for regional development often goes hand in hand with grants from the European Union's Structural Funds and Cohesion Fund, ensuring that loans and grants together bring the Bank into close collaboration with the European Commission and involve it in the preparation and implementation of structural support programmes.

The European Union is committed to a strategy for strengthening the economies of member states, their competitiveness and their capacities to create new jobs. In pursuit of these aims, the European Council has called on the Bank to play a major role in the Union's economic recovery programme.

In response, the EIB has stepped up its financing for *trans-European networks* in transport, telecommunications and energy. The Bank has set up a special lending 'window' with tailor-made financing facilities for priority projects approved by the Essen European Council in December 1994. Its long-term financing is very well suited to the particular needs of these large-scale and long-term projects.

While the European Union is the main focus of its activities, the EIB also helps to execute the financial aspects of the Union's co-operation policies with non-member states. Currently, the Bank is operating in more than 100 countries. It is supporting: economic development projects in the countries of central and eastern Europe preparing for European Union membership; cross-border infrastructure and environmental projects in the Mediterranean non-member countries; reconstruction schemes in the Lebanon, Gaza and the West Bank; technology and transfer, joint ventures and environmental protection in the Caribbean, Pacific, Asian and

Latin American countries which have signed co-operation agreements with the European Union.

Projects supported by EIB loans carry low interest rates because the Bank is able to raise the bulk of its resources from the capital markets at the best rates possible with its AAA credit rating.

The European Bank for Reconstruction and Development

The EBRD was established in 1991 to operate on a strictly commercial basis. Its objective is to foster the transition towards market-oriented economies and to promote private and entrepreneurial initiative in the countries of central and eastern Europe and the Commonwealth of Independent States (CIS). It is helping the former soviet block countries to develop into free market economies.

The EBRD aims to help its member countries to implement structural and sectoral economic reforms by promoting competition, privatisation and entrepreneurship development after taking into account the particular needs of countries at different stages of transition. Through its investments it promotes private sector activity, the strengthening of financial institutions and legal systems, and the development of the infrastructure needed to support the private sector.

In fulfilling its role as a catalyst of change, the EBRD encourages co-financing and foreign direct investment from the private and public sectors. It helps to mobilise domestic capital, and provides technical co-operation in relevant areas. It works in close co-operation with international financial institutions and other international and national organisations. In all its activities, the Bank promotes environmentally sound and sustainable development.

One of the EBRD's strengths is that it can operate in both the private and public sectors. It merges the principles and practices of merchant and development banking, providing funding for private or privatisable enterprises and for physical and financial infrastructure projects needed to support the private sector.

The EBRD aims to be flexible by using a broad range of financing instruments, tailored to specific projects. The kinds of finance it offers include loans, equity investments and guarantees.

The terms of the EBRD's funding are designed to enable it to co-operate both with other international financial institutions and with public and private financial institutions through co-financing arrangements.

By the end May 1998 the Bank had approved 574 projects involving ECU 13.9 billion of the EBRD's own funds, and these projects are expected to mobilise an additional ECU 31.2 billion. Of the approved projects 496 have been signed committing ECU 11.2 billion of the EBRD's own funds; 68% of total approved funding was for private sector projects.

Project-related technical co-operation is a major feature of the EBRD's activities. By the end of 1997, 53 co-operation fund agreements with bilateral donors, totalling ECU 512 million, had been made. This involved 1808 projects, with a total estimated cost of ECU 500 million.

The Inter-American Development Bank

The Inter-American Development Bank (IDB), the oldest and largest regional multi-lateral development institution, was established in 1959 to help accelerate economic and social development in Latin America and the Caribbean. The Bank has become a major catalyst in mobilising resources for the region. The Bank's principal functions are to utilise its own capital, funds being raised in financial markets, and other available resources to: finance the development of the borrowing member countries; supplement private investment when private capital is not available on reasonable terms and conditions; and provide technical assistance for the preparation, financing and implementation of development plans and projects. The Bank's annual lending has grown from the US $294 million in loans approved in 1961 to US $6.7 billion in 1996.

The Bank's operations cover the entire spectrum of economic and social development. In the past, Bank lending emphasised the productive sectors of agriculture and industry, the physical infrastructure sectors of energy and transportation and the social sectors of environmental and public health, education and urban development. Current lending priorities include poverty reduction and social equity, modernisation and integration, and the environment.

The financial resources of the Bank consist of the ordinary subscribed capital, reserves, funds raised through borrowing and contributions made by member countries. The Bank also has a Fund for Special Operations for lending on concessional terms for projects in countries classified as economically less developed. An additional facility, the Multi-lateral Investment Fund (MIF), was created in 1992 to help promote and accelerate investment reforms and private-sector development throughout the region. In 1994 the Bank's member countries agreed to a US $41 billion increase in the Bank's resources. Member countries' subscriptions to the Bank's capital fund consist of paid-in and callable capital. A paid-in subscription is in the form of a cash payment and represents a minor portion of a member's subscription (under the Eighth General Increase in the

Resources of the Bank, the paid-in portion represents only 2.5% of a member's subscription). The major part of a member's subscription is in the form of callable capital or guarantees of the Bank's borrowing in the world's financial markets. The Bank is able to borrow funds at very good rates with its AAA rating.

The DEG (German Investment and Development Company)

The DEG, the German Investment and Development Company, promotes private enterprises in Asia, Africa, Latin America and Central and Eastern Europe. The DEG was established in 1962. The DEG finances private investment on a long-term basis in all economic branches providing long-term loans and equity finance. The DEG also provides guarantees, which can be used to secure local loans, and arranges additional financing from other institutions. Furthermore, the DEG provides consultancy to companies already investing abroad and to those planning co-operation with foreign partners. The DEG finances business start-ups, expansion, rationalisation and modernisation investments and co-finance investments that are profitable, environmentally sound and render an effective contribution to the economic development of the host country. The DEG co-operates with German and foreign partners. The DEG's financial commitment normally ranges between DM 3 million and DM 20 million per project. There is a new trend to provide funds for smaller projects from DM 500 000 upwards.

Export finance and guarantee agencies

In order to promote export, most developed countries have set up their export finance and guarantee agencies. These agencies normally promote export by providing insurance against the country risk for the export credit obtained from banks and other financial institutions. Many of them refinance the export credit offered by the banks. They are therefore known as the *national interest lender*. The insurance normally covers the exporter's risk of loan not being paid through default of the buyer or through other causes, such as restrictions on the transfer of currency and cancellation of a valid import licence. The guarantee is provided depending on the nature of the trade.

For *repetitive trade*, which is in standard or near-standard goods and is normally of short-term nature, the insurance cover is provided under a comprehensive policy designed for the insurance of continuous business. Such guarantees normally cover 90–95% of the amount insured, depending on the scope of risks covered.

For *trade in large capital goods*, which is of a non-repetitive nature, usually of a high value and often involves lengthy credit terms, specific policies are negotiated for each contract. These policies normally cover 90% of the insured business.

Some export credit agencies also provide insurance for new investments overseas, against the political risks of expropriation, war and restrictions on remittances. Investments are underwritten individually and insurance is arranged.

Export credit guarantee premium
Premium charges on the export credit guarantee are assessed on the basis of:

- the buyer's financial position and business reputation
- the political and economic situation in the country where the buyer is located
- the period of cover.

The premium for short-term business is paid as an annual, non-refundable (holding) premium and an additional monthly premium on declaration of business. Premium rates on long-term business are determined contract by contract.

Consensus rates
In view of the cut-throat competition to export there is a tendency among the export credit agencies to offer very soft terms for export credit. This vitiates the normal market mechanism. In order to put a check on the 'credit terms race' among the OECD countries, the members of the Group on Export Credits and Credit Guarantees (ECG) of the OECD Trade Committee have reached an informal consensus that sets for most officially supported export credits of 2 years or more:

- a floor under permitted interest rates
- a ceiling on maturity
- a minimum down payment
- a maximum local cost financing allowance.

These rules have been incorporated into an Agreement on Guidelines for Officially Supported Export Credits, on 1 April 1978, in which all OECD members except Iceland and Turkey are participant. The arrangement provides that any participant who intends to offer a credit that exceeds the maximum degree of permitted

Table 10.1. Export credit guarantee agencies or national interest lenders

Country	Agency
Australia	Export Finance and Insurance Corporation (EFIC)
Austria	Oesterreichische Kontolbank Aktiengesellschaft (OKB)
Belgium	Office National de Ducroire Creditexport
Brazil	Carteris de Commércio Exterior – Banco do Brazil (CACEX) Institutio de Resegguros do Brazil (IRB)
Canada	Export Development Corporation (EDC)
Denmark	Exportkreditradet (EKR) Dansk Eksportfinansieringsfond (EF)
France	Compagnie Française d'Assurance pour le Commerce Exterieur (COFACE) Banque Française du Commerce Exterieur (BFCE)
Germany	Hermes Kreditversicherungs AG Aus Fuhrkredit-Gesellschaft mbH (AKA) Kreditanstalt fur Wiederaufbau (KFW)
Italy	Sezione Speciale per l'Assicurazione del Credito all' Esportazino (SACE) Mediocredito Centrale
Japan	Export–Import Bank of Japan Ministry of International Trade and Industry (MITI)
Korea	Export–Import Bank of Korea
The Netherlands	Nederlansche Credietverzekering Maatschappij (NCM) De Nederlansche Bank (DNB)
New Zealand	Export Guarantee Office (EXGO)
South Africa	Industruial Development Corporation of Africa Ltd (CGIC)
Spain	Compania Espanola de Seguros de Ceditio a la Exportacion (CESCE)
Sweden	Exportkreditnamnden (EKN) AB Svenska Export Kredit (SEK)
Switzerland	Exportrisikogarantie (ERG)
Taiwan	The Export–Import Bank of China
UK	Export Credit Guarantee Department (ECGD)
USA	Export–Import Bank of USA Private Export Funding Corporation (PEFCO) Overseas Private Investment Corporation (OPIC)

concessionality should notify that intention beforehand to other participants and explain the reasons for the intended action. Participants meet at the OECD headquarters for an annual review of the functioning of this arrangement and its appropriateness.

Important export credit guarantee agencies

Some of the important export credit guarantee agencies are listed in Table 10.1.

The Overseas Private Investment Corporation

The Overseas Private Investment Corporation (OPIC) also insures projects against political risk. In order to be eligible for risk insurance by the OPIC, at least one of the projects investors must be US based, as the OPEC is funded by the US government. The only private insurer of importance against expropriation risks is Lloyd's of London.

Bibliography

Baum, W. C. and Tolbert, S. M. *Investing in Development-Lessons of World Bank Experience*. Oxford University Press, New York, 1985.

Beasant-Jones, J. A view of multi-lateral financing from a funding agency. The World Bank. Paper presented at: *Financing Hydropower Projects – 1994*, 22–23 September 1994, Frankfurt.

Boughton, J. M. and Lateef, K. S. *Fifty Years After Bretton Woods: The Future of the IMF and the World Bank*. International Monetary Fund and The World Bank Group, Washington, DC, 1995.

Carter, L. W. and Bond, G. *Financing Private Infrastructure Projects: Emerging Trends from IFC's Experience*. Discussion Paper No. 23 (1994, updated version 1996), International Finance Corporation, Washington, DC.

Department for International Development. *Eliminating World Poverty: A Challenge for the 21st Century*. HMSO, London, 1997.

Ferreira, D. and Khatami, K. *Financing Private Infrastructure in Developing Countries*. Discussion Paper No. 343, World Bank, Washington, DC, 1996.

French, D. and Sward, H. *A Dictionary of Management*, revised edition. Pan Books, London, 1984.

George, S and Fabrizio, S. *Faith and Credit, The World Bank's Secular Empire*. Penguin, London, 1994.

Giuseppe, S. *International Organisations: A Dictionary and Directory*. London.

Harvey, C. *Analysis of Project Finance in Developing Countries*. Heinemann, London, 1983.

International Finance Corporation. *Emerging Stock Markets Factbook, 1995*. IFC, Washington, DC, 1995.

Merna, T. and Njiru, C. *Financing and Managing Infrastructure Projects*. Asia Law and Practice, Hong Kong, 1998.

Nevitt, P. K. *Project Finance*, 4th edition. Bank of America, Financial Services Division, 1983.

Nicholas, H. G. *The United Nations as a Political Institution*, 5th edition. Oxford University Press, Oxford, 1975.

Piesse, Peasnell, K. and Ward, C. *British Financial Markets and Institutions – An International Perspective*, 2nd edition. Prentice Hall, London, 1995.

Richardson, R. W. and Haraiz, J. H. *Moving to the Market: the World Bank in Transition*. World Bank, Washington, DC, 1990.

Selim, H. M. *Development Assistance Policies and the Performance of Aid Agencies*. 1983.

Taylor, P. and Groom, A. J. R. *International Institutions at Work*. Cassell, London, 1988.

Wells, S. J. *International Economics*. Minerva, London, 1973.

World Bank. *World Development Report, 1994: Infrastructure For Development*. Oxford University Press, Oxford, 1994.

World Bank. *The World Bank Guarantee: Catalyst for Private Capital Flows*. Project Finance and Guarantees Group, Resource Mobilisation and Cofinancing, World Bank, Washington, DC/Oxford University Press, 1996.

CHAPTER ELEVEN

Privatisation as a method of financing infrastructure projects

Introduction

The challenge of providing infrastructure services has led governments in both developed and developing countries all over the world to look at new and innovative alternative methods of financing and managing infrastructure projects. The most viable alternative method is through privatisation. *Privatisation* refers to a shift from publicly to privately produced goods and services via the sale of public assets, infrastructure or enterprises.

Definition and justification of privatisation

Gayle and Goodrich (1990) have defined privatisation as the process of reducing the roles of government while increasing those of the private sector in activities or asset ownership. Mclindon has stated that privatisation and capital market development are key to reform and development since they mitigate the distortions that follow from flawed strategies and promote economic growth through several different but complementary channels. Privatisation plays a pivotal role in economic reform. It enables a government to shift its portfolio of economic interventions out of areas of the economy inwhich the private sector is able to operate more efficiently and productively. This frees resources for those areas that are the basic responsibilities of government. Studies have shown that privatisation improves the competitiveness and efficiency of enterprises, which promotes economic growth.

Privatisation should ideally be preceded by liberalisation in order to open up the economy to international competition. Privatisation should then be carried out in a transparent and accountable manner, followed by deregulation in order to make the privatised enterprises face market forces. *Deregulation* refers to the disengagement of government from specific kinds of responsibilities. *Liberalisation* refers to the opening up of any industry to competitive pressures. This is done through removal of government regulation of the market.

The expected economic benefits of privatisation include:

- increased quantity of production
- improved quality of the output
- reduced unit cost of production
- expanded opportunities for growth and employment in the longer term
- generation of new technologies
- increased foreign investment.

When accompanied by liberalisation, privatisation encourages the emergence of managers who are willing to champion an entrepreneurial risk-taking culture. Corporations then become more results oriented, displaying such new attributes as aggressive marketing styles, improved management information systems, and reduced overhead costs. In principle, the empirical evidence for the relative economic efficiency of the private sector, as opposed to public sector production, is overwhelming.

It has been argued that the role of governments is to ensure adequate investment in people, provide a competitive climate for private enterprises, keep the economy open to international trade, and maintain a stable macro-economy. It has often been observed that when governments try to do more they are likely to do more harm than good. Attempts to guide resource allocation with non-market mechanisms have generally failed to improve economic performance. The challenge of meeting the rising demand for infrastructure at a time when governments have limited resources has led to the realisation that the government's portfolio of interventions in the economy should be shifted. Privatisation provides a way of achieving the desired shift from public to private provision of goods and services.

Types of privatisation

Privatisation has been defined previously as the act of reducing the role of government, or increasing the role of the private sector, in an activity or in the ownership of assets. Governments pursuing privatisation can choose from a number of options. In practice, different types of privatisation are often used in combination. For instance, the government may sell some large stakes in a state-owned enterprise to a local or foreign investor, who will manage and control the company and bring in new capital, technology and market access. Smaller percentages of the shares are offered to employees of the enterprise, while the public may also be offered shares in a public share offer.

The basic types of privatisation include:

- liquidation
- contracting out
- leasing
- deregulation and demonopolisation
- management–employee buy-outs
- trade sales
- public share sales
- mass privatisation
- build–own–operate–transfer (BOOT) infrastructure privatisation.

Each of the main types of privatisation is considered in some detail below.

Liquidation

One privatisation measure is simply to liquidate some state-owned enterprises. This may be warranted in cases where no combination of new investment, ownership and operational changes exists which would give the enterprise a positive net present value in terms of future cash flows.

Contracting out

A number of services, including management, can be contracted out to the private sector, usually through a competitive bidding process followed by negotiations. Contracting out is a useful technique for some services, such as refuse collection and disposal, design and site supervision services from consultants, and construction from contractors. Contracting out services usually leads to an improvement in day to day operations. Contracting out does not, however, address the need for new investment in plant, equipment and technology, which keeps an enterprise competitive. Another drawback is that contracting out does not transfer ownership to the private sector, which is critical for sustained improvement in enterprise performance.

Leasing

A lease enables a private sector group to control a company, assets, or both for a period of time, for financial gain. Leasing is a more powerful tool than contracting out, since the returns to the lessee are more directly determined by the lessee's success in managing the leased assets. As with contracting out, however, the problem of making new investment in the enterprise, and the problem of ownership, remain. This method has been applied widely in economic infrastructure, such as water supply systems where a private operator leases water supply infrastructure for a period of time and relieves the public water organisation of the daily operation of the water supply.

Deregulation and demonopolisation

Under this form of privatisation the government removes regulations, which had previously prevented the private sector from competing with a state-owned monopoly, or changes old regulations or creates new ones to foster greater competition from the private sector.

Management–employee buy-outs

Occasionally, management, employees, or both may be ideal buyers for a state-owned enterprise, especially for smaller enterprises. This option is attractive when management is effective but ministerial directives and other operational constraints have compromised performance. *Leverage buy-outs* (LBOs) have played a role in arranging finance for such buy-outs.

Trade sales

A trade sale involves the sale of controlling interests in an enterprise to the private sector, usually through competitive tendering techniques, an evaluation of proposals, and negotiation between the government and the potential private sector buyer.

Public share sales

Public share sales can overcome the potential problems of trade sales by virtue of their greater transparency and openness to the public. Properly structured, marketed and priced, broad-based public share sales can help to popularise the privatisation programme, spread the benefits of economic reform and ownership, and develop capital markets. Public share sales were the most salient form of privatisation in the 1980s and 1990s.

Mass privatisation

Mass privatisation is a new and rapid approach that enables countries to privatise thousands of enterprises, utilising standard techniques, transparent procedures, and vouchers distributed to citizens. Mass privatisation through vouchers has expanded the envelope of privatisation techniques and strategies, and proven that substantial privatisation can take place even in the absence of capital markets, adequate accounting standards, and legal and other infrastructure, all of which were formerly considered prerequisites to privatisation.

McLindon (1996) states that mass privatisation, when properly structured, can serve as a catalyst for the development of capital markets where none previously existed, although complementary measures are needed to safeguard the gains from mass privatisation. He has given the Czech and Slovak Federal Republic, Russia, Lithuania, Moldova and Kyrgyzstan as examples of countries that have initiated successful mass privatisation programmes with vouchers and simultaneously initiated capital markets.

BOOT infrastructure privatisation

The BOOT model is a method of privatisation well suited for addressing the needs of infrastructure. This type of privatisation involves economic infrastructure (e.g. power, roads, ports, water supply, wastewater collection, treatment and disposal) that it would be desirable to construct but cannot be built owing to budgetary limitations. Using BOOT and related techniques, the government tenders these projects to the private sector, which assumes responsibility for the financing, construction and operation of the infrastructure project. BOOT infrastructure privatisation, which is becoming more prominent in both developed and developing countries, will draw on capital markets for finance, and in turn will spur significant capital market development.

The BOOT method of privatisation of infrastructure is a sustainable method of private sector participation in the financing and management of infrastructure projects. Details of this financing strategy are provided in Chapter 12.

Goals of privatisation and deregulation

The prime goals of privatisation are to:

- increase individual choice
- balance budgets
- reduce debt
- raise capital (including foreign exchange)
- spread popular capitalism.

The prime objectives of deregulation are to:

- eliminate dysfunctional government regulation
- stimulate productive efficiency
- encourage competition
- prompt corporate self-regulation.

Prerequisites for privatisation

Privatisation and deregulation are policies that excite controversy because they imply fundamental and differentially motivated changes in the boundary between state and society. Some kinds of privatisation and deregulation raise distinctive concerns. For instance, in the case of contracting, but not divestiture, issues such as accountability and quality assurance are crucial. There are several common requisites and considerations.

The policies require a stable political consensus that the governing elite will become more, not less, influential as a result. The dilemma as to whether to emphasise maximum possible price or widest practicable share ownership in the course of public offers for sale reflects this imperative. Most governments remain reluctant to cede control (as opposed to ownership) over public enterprises, even though privatisation can increase state control over price and service quality objectives.

The policies are premised on a recurrently observed search for enhanced economic efficiency by means of selective market expansion and liberalisation. The benefits of privatisation are best achieved when there is market liberalisation. It is, however, possible to increase competition without privatisation. Inclusiveness is important for successful privatisation. This requires co-opting opposing workers, managers and consumers.

Benefits of privatisation

Privatisation is beneficial to any economy, both in developed and developing countries. The main benefits of privatisation are:

- greater efficiency and productivity of the enterprise
- generation of revenue to reduce deficits and debt
- capital market development
- attracting foreign investment.

These benefits of privatisation are considered in some detail below.

Privatised enterprises can focus on being competitive to produce, at low cost and acceptable quality, the goods and services that consumers want and are willing to pay for. This would lead to a more efficient use of resources and improve overall economic output. There is empirical evidence that beneficial changes follow privatisation. These include an increase in investment, a rise in prices toward levels that reflect scarcity values, greater productivity due to managerial effort, better marketing and diversification, and freedom to shed excess labour.

A broader study carried out to compare the pre- and post-privatisation performance of 61 companies in both developed and developing countries indicated that there were significant increases among newly private firms in profitability, output per employee, capital spending and employment. This study showed that profitability, efficiency, investment, output and employment increased after privatisation. The study's most surprising conclusion was that employment in all the enterprises actually increased by 6% following privatisation. The study showed that debt/equity ratios improved after privatisation and that dividends as a percentage of sales increased.

It should be noted, however, that the efficiency and welfare gains from privatisation of integrated monopolies such as infrastructure needs a closer look. Privatisation of integrated monopolies alone does not automatically increase competition. Governments should consider *unbundling* parts of the monopoly, i.e. identifying the segments of the monopoly that are potentially competitive (e.g. power generation for a power company) and promoting new entry and competition by the private sector in these segments. An integrated monopoly will generally be more attractive to investors than an unbundled monopoly. In such a case the government should set up a regulatory authority to regulate the privatised monopoly in order to derive the full benefits of privatisation.

Government budgets worldwide are stretched to the limit. Most governments around the world can only meet bills by raising taxes or by inflating the currency to reduce the real value of debt. Since taxes in most countries are already too high, selling state-owned enterprises through privatisation is a viable way to augment government budgets. Privatisation generates revenue from the sale of shares in state-owned enterprises, eliminates the need to provide subsidies, and increases tax revenues from restructured and more productive enterprises. All these factors help restore fiscal balance and relieve inflationary pressures.

Privatisation is particularly beneficial in developing countries as it can ease a country's foreign debt burden. State-owned enterprises have been some of the main beneficiaries of government intervention, especially in terms of foreign debt. The World Bank's 1989 *World Development Report* noted that by 1989

> ... the outstanding stock of foreign loans to state-owned enterprises for a sample of 99 developing countries was twice that to the private sector. Borrowing was necessary not just for investment but also to cover losses.

Lower deficits, achieved with the help of privatisation revenue, reduces the government's need to borrow, which eases the 'crowding out' of the private sector from financial markets. With privatisation proceeds there could be reduced taxation, increased disposable incomes and increased savings by households that, when coupled with reduced government demand for savings, tends to reduce interest rates.

Lower interest rates help all private sector firms, and can lead to an increase in share price for those firms that are listed. Before privatisation, some state-owned enterprises require huge subsidies, which lead to printing press financing and inflation. After privatisation, these enterprises face hard budget constraints and limits on credit from the banking system. This is crucial for reducing inflation and the uncertainty of financial transactions. Once restructured, many enterprises become profitable and can contribute tax revenues to the government budget.

Privatisation leads to increased efficiency, quality of service and profitability of the enterprise, among other benefits to the economy. Privatisation could lead to reduced taxation, as observed in the UK where the base rate of personal income tax (direct taxation) reduced from 33% to 20% in 1979 and 1998, respectively, as a result of privatisation of state-owned enterprises.

One of the most important benefits of privatisation is its impact on capital markets, which in turn can facilitate and contribute to savings, investment and economic growth. Privatisation has a macroeconomic impact on the development of capital markets. Privatisation of state-owned enterprises and economic infrastructure reduces deficits and inflationary pressure, which builds a stronger foundation for capital markets. Privatisation can be a means of deepening domestic capital markets. Public sale of shares, mass privatisation and build–own–operate (BOO) or build–operate–transfer (BOT) privatisation lead to the creation of tradable securities. Public share sales and mass privatisation help create broad and diversified share ownership, new companies listed on the stock exchange, and new investment funds. The selling of state-owned enterprises increases market capitalisation and liquidity, as well as adding stability to the stock market. Privatisation by share sale can help to transfer the financial technology of *initial public offerings* (IPOs) to the fledgling local securities industry, and have a demonstration effect by encouraging private sector companies to undertake their own IPOs and secondary offers to raise equity financing. These are new vehicles to channel savings into productive investment, which is key to economic growth. Privatisation can also be decisive in developing local institutional investors, such as mutual funds, insurance companies and pension funds, which are critical to expanding capital markets and making them professional.

Developing countries are finding it increasingly important to attract foreign capital to achieve higher rates of growth. Without foreign capital, the investment that fuels growth will be limited to domestic savings. It is unlikely that the financing role of foreign bank loans to those countries that need it most will return to the high levels of the 1970s. Official development assistance, especially at a time of fiscal conservatism by the Organisation for Economic Co-operation and Development (OECD) countries, cannot be expected to fill the gap. The additional foreign capital must, therefore, come through foreign direct and portfolio investment. Many countries that are privatising would like to attract a strategic foreign investor into a state-owned enterprise because such investors can bring capital, new technology, new export market access and more independent professional management to the enterprise. Privatisation, by BOOT techniques in particular, attracts foreign direct investment. The huge levels of investment needed in economic infrastructure have encouraged governments to permit foreign and local private sector investors to bid on infrastructure projects that otherwise might not be built owing to budgetary limitations.

Privatisation also encourages foreign portfolio investment. Portfolio investors in developed countries are attracted by the high returns in emerging markets and low correlation with developed markets, which lowers their overall portfolio risk. A plethora of investment funds has developed to give western investors the opportunity to access the capital markets in developing countries. These funds include general emerging market funds, regional funds, specific country funds, telecommunications funds and infrastructure funds. Some funds specialise in purchasing the shares of companies being privatised. Managers of these funds have directed billions of dollars to emerging markets.

Foreign portfolio investment in emerging markets increases demand for equities. This demand, other things being equal, will lead to stronger share prices, which will lower the cost of capital to companies to issue new shares. In this way, foreign portfolio investment encourages new issues and new listings in developing countries.

Foreign portfolio investors are an additional source of demand for state-owned enterprise shares offered for privatisation. Governments may limit participation in privatisation to local residents in order to avoid criticism that state-owned enterprises are being sold off to foreigners.

Foreign portfolio investors make the markets more efficient by using analytical techniques to buy undervalued shares and sell overvalued ones. They usually demand better information than domestic investors do, which may improve local investment research abilities, accounting standards, audit procedures and disclosure requirements.

Foreign portfolio investors have been instrumental in the emergence of emerging markets over the last decade. In 1985, net portfolio equity flows were $150 million to developing countries, and in 1993 they were $46.9 billion, which was nearly as much as the official development assistance of $53.9 billion.

World trends in privatisation and economic reform

There is growing worldwide interest in reforms to limit government spending, reduce taxes which discourage investment and employment, cut subsidies, remove barriers to trade and investment, eliminate distortions, develop capital markets and privatise.

The UK was a pioneer in privatisation in the 1980s. Its policy towards economic reforms is similar to that articulated by the international donor agencies. A recent policy document from the UK Department for International Development (1997) states that:

> We intend to develop a new partnership with the private sector in Britain. We want to work with British businesses, since long-term trade and investment is essential to stimulate the growth, which benefits everyone, especially those most in need. In future, developing countries are going to be increasingly important markets for British goods and services, so it makes business sense as well as moral sense for British businesses to be involved. As part of this new partnership we will, for example, provide information to British companies about trade and investment opportunities in developing countries, make sure that development projects make full use of British business skills, and work with British business networks in developing countries.

Trade and investment are vital for eliminating poverty. They help generate growth, which in turn helps people get jobs and earn a livelihood.

Privatisation in developing countries

Most countries in the world pursued a development strategy in which the state-owned enterprises played a leading role, and have experienced poor results. A recent World Bank study (1989) found that in many developing countries with large state-owned enterprise sectors:

... the inefficiency of the state-owned firms, combined with the attendant state enterprise sector deficits, are hindering economic growth, and making it more difficult for people to lift themselves out of poverty.

The study found out that the financing of state-owned enterprises is a burden on government finances and developing country banking systems, and may undermine fiscal stability and fuel inflation.

The harsh economic situation in developing countries has necessitated changes in all spheres of the economy. Most developing countries are changing their economic policies in order to survive in the modern world. The benefits of privatisation have been recognised by many developing countries that have embraced privatisation and other economic reforms.

Bibliography

Galal, A. and Shirley, M. (eds). *Does Privatisation Deliver?* World Bank, Washington, DC, 1994.

Gayle, D. J. and Goodrich, J. N. *Privatisation and Deregulation in Global Perspective.* Printer, London, 1990.

International Finance Corporation. *Emerging Stock Markets Factbook, 1995.* International Finance Corporation, Washington, DC, 1995.

McLindon, M. P. *Privatisation and Capital Market Development, Strategies to promote Economic Growth.* Praeger, USA, 1996.

Megginson, W. L., Nash, R. C. and van Randenborgh, M. *The Privatisation Dividend: A World-wide Analysis of the Financial and Operating Performance of Newly Privatised Firms.* Public Policy for the Private Sector, World Bank Group, Washington, DC, 1995.

Merna, T. and Njiru, C. *Financing and Managing Infrastructure Projects.* Asia Law and Practice, Hong Kong, 1998.

UK Department for International Development. *Eliminating World Poverty: A Challenge for the 21st Century.* HMSO, London, 1997.

World Bank. *World Development Report, 1989.* Oxford University Press, New York, 1989.

World Bank. Summary. *Bureaucrats in Business.* Oxford University Press, New York, 1995.

CHAPTER TWELVE

Typical risks in the procurement of infrastructure projects

Introduction

Many projects are considered high risk because of the amount of uncertainty involved in the method of financing them. It is necessary to increase the chance of a project succeeding by identifying the risk associated with the finance and taking the necessary actions. The risks associated with financing major projects, whether by public or private organisations, are numerous. All client organisations need to consider the risks typically associated with major projects, these being construction and operation risks, the risks associated with revenue generation and risks associated with how the project is to be financed.

The risks associated with the method of repayment are the uncertainty of cash inflows and outflows. Many projects need flexible repayment mechanisms to ensure success should changes in cash flows occur. Project cash flows should be considered for worst- and best-case scenarios, to ensure that repayments can be made when risks occur. Some of the most important risks to consider are financing risks.

Appraisal and validity of financing projects

The financial viability of a project must be clearly demonstrable to potential investors and lending organisations. In assessing the attractiveness of a financial package, project sponsors should examine the risks associated with the project.

The three basic financial criteria that need to be achieved in projects are:

- finance must be cost-effective, so far as is possible
- the skilled use of finance at fixed rates to minimise risks should be adopted
- finance should be acquired over a long term, thus eliminating refinancing risk.

The project must have clear and defined revenues that will be sufficient to service principal and interest payments on the project debt over the term of the loans and to provide a return on equity which is commensurate with development and long-term project risk taken by equity investors.

When selecting the sources and forms of capital required, the strength of the security package, perception of the country risks and limits and the sophistication of local capital markets should be considered.

One of the most important elements to be satisfied in a project is how to provide security to non-recourse or limited recourse lenders. If a promoter defaults under a project strategy utilising a non-recourse finance package, the lender may be left with a partly completed facility that has no market value. Various security devices are often included to protect lenders, and these are:

- revenues are collected in one or more escrow accounts maintained by an escrow agent independent of the promoter company
- the benefits of various contracts entered into by the promoter (e.g. construction contract, performance bonds, supplier warranties, insurance proceeds) will normally be assigned to a trustee for the benefit of the lender
- lenders may insist upon the right to take over the project (step-in clauses) in case of financial or technical default prior to bankruptcy and bring in new contractors, suppliers or operators to complete the project
- lenders and export credit agencies may insist on measures of government support such as stand-by subordinated loan facilities which are functionally almost equivalent to sovereign guarantees.

The successful elements that are required in funding projects could include:

- limited and non-recourse credit
- debt financing entirely in local currency
- equity finance in currencies considered relatively strong
- confident project creditors and governments prepared to accept some project risks and provide limited resources.

The contract between the client and lender can only be determined when the lender has sufficient information to assess the viability of a project. In most projects the lender will look to the project itself as a source of repayment rather than the assets of the project. The key parameters to be considered by lenders include:

- *Total size of the project*: this determines the amount of money required and the effort needed to raise the capital, the internal rate of return on the project and the equity.
- *Break-even dates*: critical dates when equity investors see a return on their investments.
- *Milestones*: significant dates related to the financing of the project.
- *Loan summary*: the true cost of each loan, the amount drawn and the year in which draw downs reach their maximum.

A properly structured financial loan package should achieve the following basic objectives:

- maximise long-term debt
- maximise fixed rate financing
- minimise refinancing risk.

It is important to realise that the financial plan may have a greater impact on the terms of a project than will the physical design or construction costs.

Managing financial risks

Both borrowers and lenders need to adopt a risk management programme. Risk management should not be approached in an *ad hoc* manner, but should be structured. The five major steps of such a process are to:

- identify the financial objectives of the project
- identify the source of the risk exposure
- quantify the exposure
- assess the impact of the exposure on business and financial strategy
- respond to the exposure.

Financial objectives

It is important to understand the financial objectives for a project as a basis of managing financial risks. The first stage is to develop a clear understanding of the project. Borrowers and lenders need to determine their objectives regarding the financing of a project. Many borrowers seek long-term loans with repayments made from revenues. The risk of not meeting repayments is often reduced when the borrower has sufficient earnings at the start of operation to service the debt. Many projects, however, suffer commissioning delays that increase the borrower's loans and repayments. In many cases borrowers will seek grace periods from lenders to cover such delays.

Lenders seek positive cash flows and must ensure that their objectives are met by providing the best loan package. If a short-term loan is the lender's objective then the major risk will occur at the start of operation, and should the project not generate sufficient revenues the lender may, for example, consider debt for equity swaps.

Once the project objectives have been defined and the overall costs, including construction and operation costs, determined, a cumulative cash flow model is prepared. The model can be used to quickly estimate the *net present value* (NPV), the *internal rate of return* (IRR) and the pay-back period of a project. This model is initially prepared without considering potential risks. It is essential that the estimates and programmes prepared are reflective of cost and time over the life cycle of the project. The risk of inaccurate estimates based on fixed budgets often leads to optimistic cash flows that do not truly illustrate the effects of risk occurring during a project.

In many cases the cost of finance is not included in the cash flow at this initial stage. Many organisations prefer to use the *return on investment* (ROI) as the measure of profitability. The authors, however, consider that the cost of finance along with all other projected costs and revenues should be incorporated into the cash flow, as this provides a more accurate illustration of the project's finances. The working capital should also be considered in the cash flow since certain risks may occur and result in further borrowing over and above that estimated.

Identification of project risks

Risks can be identified specific to one project or on a portfolio of projects. The risks may be identified on the basis of historical market data or past experience of similar projects. It should be noted that identification of risks may be beneficial to a project and that competitive advantages can be gained by organisations who consider such risks at project appraisal stage. All the risks identified, especially financial risks, should be built into the project cash flow at this stage. It is also important to consider the effect of risk with time, as with many projects certain risks decrease with time while others increase.

In most projects, borrowers and lenders are concerned primarily with three risks:

- cash flows
- interest rates
- market risks.

In this chapter the three major risks are used to illustrate the risk management process.

Quantification of risks

Having identified a number of risks, the effect of the risks on the overall commercial viability of a project can be quantified. At this stage the sources and types of financial instruments will provide the basis of the financial estimates.

If, for example, the risks of an increased expenditure in construction, an increase in base rate interest and a reduction in market output are identified, then the cost associated with each risk can be determined in terms of actual money spent or lost. In this example, it is assumed that there are no changes in the project time scale.

This may result in a 5% increase in construction costs, a 3% increase in the amount of money to be repaid and a 10% reduction in the revenues generated.

The effect of these risks can then be illustrated in terms of NPV, IRR and pay-back period. If the objective is to pay back the loan as soon as possible then the borrower must consider the costs of meeting such requirements by covering each risk.

A cumulative cash flow diagram can be prepared representing the effect of the risks and compared with the original cash flow diagram.

Assessment of the impact of risk exposure

Both borrowers and lenders must assess the effects on the commercial viability of the project once the risks have been quantified. In some cases the borrower may use the *minimum acceptable rate of return* (MARR) of a project as the basis for sanction. Lenders may consider that the project is too risky to finance because of the sensitivity of the market. In other cases a project may be sanctioned or rejected by comparing it with returns that could be secured on a similar, but less risky, project.

At this stage borrowers and lenders will consider the types of financial instruments to be used to determine the cash flow. This exercise may be repeated using different financial instruments. Tailoring loans to suit the project can often be used to alleviate many of the risks identified. This is often referred to as *financial engineering.*

Response to risks

At this stage the selection of appropriate risk management products is considered.

Increase in construction costs

An increase in construction costs may be overcome by a number of responses, such as:

- applying a fixed price lump sum contract
- changes in design to reduce costs
- constructing the project for early on-stream production.

Each of the above can be quantified in terms of money, time and the cost of response. For example, the cost of a lump sum fixed price construction contract may cost 3% more that the costs assumed in the initial estimate. At this stage borrowers and lenders must consider whether the additional cost of this form of contract is worth the risks covered.

Similarly, the cost of changing the design and specification may result in a lower construction cost. However, this must be balanced against the possibility of a reduction in quality or higher operating costs.

In some cases early product off-take can provide revenues to reduce borrowing. However, this may result in the necessity to re-programme the project and incur additional construction and operation costs.

Increase in the rate of interest

An increase in the rate of interest may be overcome by:

- the use of caps and floors
- the use of a fixed interest rate.

A *cap* is a ceiling or upper bound on a specified interest rate, and a *floor* is simply a lower bound on a rate. The borrower will pay for a cap and be paid to take a floor.

Changes in interest rates over the project life cycle can seriously affect the commercial viability of a project. Market uncertainty, political decisions and inflation affect interest rates and their effect on the project must be considered.

If the project can bear a 2% increase in interest rates over the first 5 years of operation then the borrower may wish to retain that risk. If, however, an increase of over 2% would make the borrower unable to meet debt service, then a cap should be purchased. Similarly, the lender will seek to achieve the best interest on the loan and pay the borrower to accept a floor. The authors suggest that when interest rates are low, the borrower should seek a fixed interest rate that will meet the requirements of debt service. If, however, interest rates are high then the borrower should seek a floating rate, in the hope that rates will be reduced and a fixed rate can be set.

Market risk

Market risks are probably the most uncertain risks to be considered for a project. Many projects may be commercially viable provided revenues remain as forecast. However, changes in the market, such as demand, obsolete product or new technology, may occur. Re-investment of profits back into a project does not always cover such risks.

The borrower must consider the effect of such risks and how best to respond to them. Response may be in the form of insurance, forward contracts or sale of the asset. Again the borrower must consider the cost of premiums associated with covering such risks. It is important to mention that a lot of insurance cover does not make a good project out of a bad one, and the premiums required must be compared with the cost of risks covered.

Finally, the cumulative cash flow diagram can be prepared to accommodate the cost of response to risks and the commercial viability of the project can then be determined. In this section only the effect of financial risks has been considered. The authors suggest the same format for risk analysis to cover all risks, as they can all be converted to costs and revenues.

Financial risk

The word 'risk' is normally associated with an unexpected and undesirable change. This is, however, not always true for financial transactions. In financial transactions there are always two parties who hold diametrically opposite viewpoints. Consider a project that decides to fund part of its capital requirements through an issue of fixed rate coupon bonds. Suppose that after the issue has been made the market interest rates goes up. The issuer of the bond will have an occasion for jubilation because it will continue to pay a lower coupon rate to the investors. However, the investors in the bond will view it as an undesirable situation because they will continue to receive the already fixed coupon rate that is lower than the prevailing interest rates. The same change (event) is desirable for one party and undesirable for another. On the other hand, had it been a floating rate bond issue, the issuer of the bond would have considered the increase in market interest rates as undesirable and the investors in the bond would have considered it desirable. Since in a financial transaction there are always two parties, a particular change can affect the two parties in diametrically opposite ways. It is therefore more prudent to define risk as 'any variation in an outcome' (Galitz, 1995).

Sources of financial risk

Financial risk may be defined as the impact on the financial performance of any entity exposed to risk. This definition makes it clear that any event or action, which has an impact on the financial performance of an entity, is a financial risk. This is, however, a very broad definition and it is really difficult to prepare a foolproof list of all possible events or acts that may have an impact on the financial performance of an entity. The principal sources of financial risk may be broadly classified under the following major headings (Merna, 1998):

- currency risk
- interest rate risk
- equity risk
- commercial risk
- liquidity risk
- counterparty risk
- country risk or political risk
- accounting risk and economic risk.

Currency risk

Currency risk arises when there is cross border flow of funds. With the collapse of fixed parities in the early 1970s, exchange rates of currencies are free to fluctuate according to the supply and demand for different currencies. The operation of speculators in the money market has added to the volatility of the exchange rates. Foreign exchange transactions involving any currency is, therefore, subject to currency risk. A *convertible currency* is one that can be freely exchanged for other currencies or gold without special authorisation from the appropriate central bank. Currencies of the developing countries are more prone to this risk because of the generally high levels of inflation in these countries. *Exotic currencies* are those for which there is no active exchange market. Most of the currencies of the under-developed world fall within this category.

A large number of projects, whether privately funded or funded by multi-lateral or bilateral agencies, draw on foreign capital and therefore face the risk of movements in exchange rates. International lenders very rarely assume this risk. They mostly insist on denominating their repayments in hard currency, usually the currency of the country of the lending organisation. In the past, public enterprises or governments have borne the currency risk. With growing privatisation, the currency risk now falls on the projects and, ultimately, on the consumers of the services. In many build–

own–operate–transfer (BOOT) projects the pricing of services is linked to an international currency.

Interest rate risk

The term 'interest rate risk' implies an exposure to movements in interest rates. It directly affects both the borrowing and the investing entity. The exposure depends on the maturity of the funds raised and the developments in the financial market from where funds have been raised.

Interest rate risk can broadly be classified in two categories:

* The risk on securities or financial instruments used for raising short-term finances. These facilities (e.g. Commercial Paper, Banker's Acceptance Facility) mature over a short period. The interest rate risk on these facilities depends largely on the developments in the money market.
* The risk on financial instruments that have a longer maturity but where the longer period is split into smaller periods. An example is a floating rate 5-year borrowing where the interest rate is fixed every 6 months and the borrower is exposed to different rates every 6 months until the loan matures.

Equity risk

A rise or fall in share prices affects the entity who is holding the instrument. However, it also affects companies whose shares are publicly quoted. Such companies may face difficulty in raising finance if the market price of their shares falls significantly in value. Equity risk becomes more pronounced when a borrower has issued warrants and convertibles.

If a company issues warrants that give the warrant holder the right, but not the obligation, to buy a specified number of ordinary shares from the company at a fixed price during a given period of time and at a specified price, then the company expects that the warrants will be utilised and the company will be able to receive funds in lieu of converting warrants into shares. The chances that the warrant will be utilised by the warrant holder depends on the market price of the shares. If the company has not been performing well there is little chance that the warrant will be exercised and the company receive funds.

Similarly, a convertible bond gives the holder of the bond the right to exchange it for a given number of shares before maturity of the bond, and a convertible preference share gives the holder a right to change it to an ordinary share without any time limit specified. Changing the instrument from debt to equity also changes the gearing of the company. When the company is not performing well it will prefer a low gearing, but the holders of the convertibles would like to retain the bond and not change it to equity because that may reduce the returns on their investment.

Commercial risk

There are a large number of commercial risks that can affect the financial performance of a project company. We can broadly classify them as:

- risk relating to the completion of the project
- risk relating to the operation of the project
- risk relating to the input and output of the project.

The risks relating to the completion of the project may involve the risk that the construction of a project will not be completed within the required time frame as a result of, say, contractor delays. It may involve cost overruns because of delays, or an underestimation of the costs involved. Risk of completion may also manifest in the project failing to be commissioned with the required performance specifications.

The risks relating to the operation may be that the project does not operate with the desired efficiency, that the project operation is more costly than projected, or that the project gets involved in certain legal issues such as environmental liabilities or a natural calamity such as a fire.

The risk relating to the inputs and outputs arises because every project depends on certain supplies (resources) to produce something else, which is in demand by some other entity – the supply side risk and the demand side risk. The *supply side risk* exposes the cost of production and the *demand side risk* exposes the revenues of the company. Supply side risk may involve an inadequate or inconsistent supply of raw material or other inputs (e.g. supply of gas to a gas based power company by another agency, private or governmental) or a price increase by the input suppliers.

On the demand side, the risk could be an inadequate demand for the project output (either in terms of price per unit or in terms of quantity). In countries where more and more infrastructure development sectors are being opened up to the private sector, the success of a project depends on the credibility and solvency of the buyers of the project. For example, in a private sector power project the offtake of the power may be by a government utility that transmits and distributes power. If this utility company suffers from certain bottlenecks of inefficiency or an inadequate supply network then the project runs the risk of 'offtake' by the power transmission and distribution agency.

Liquidity risk

Liquidity risk, in fact, is an outcome of the commercial risk. If the project is not able to generate sufficient resources to meet its liabilities it enters into liquidity risk. Liquidity risk is the potential risk arising when an entity cannot meet payments when they fall due. It may involve borrowing at an excessive rate of interest, or facing penalty payments under contractual terms, or selling assets at below market prices (this is sometimes classified as *forced-sale risk*). Liquidity risk is extremely important because most of the borrowing, whether a loan or a bond, has a *cross default clause*. This means that if the company has defaulted on any of its obligations then a debt with a cross default clause may be called back by the lenders for immediate repayment even if it is not due for repayment. If this provision is triggered then the company may face even more liquidity problems and may be forced to declare bankruptcy.

Counterparty risk

Any financial transaction involves two parties. Both the parties run the potential risk of default by the other party. An alternative name for counterparty risk is *credit risk*. For example, if a company has tied a line of credit from a bank or financial institution then it runs the risk of the lender not being able to meet its commitments in providing the funds at the right time. On the other hand, after the loan has been disbursed the lender runs the risk of default in repayment and interest payment by the borrower. The magnitude of the counterparty risk depends on:

> ... the size of all outstanding positions with a particular Counterparty and whether or not any netting arrangement is in force. (Galitz, 1995)

Sometimes counter party risk is differentiated from replacement risk. *Replacement risk* is on account of the potential loss if a transaction has to be replaced prior to its maturity date. Derivative market instruments such as forward rate agreements (FRAs), futures and options involve less risk than do cash market transactions, because the principal amounts are not generally exchanged. The existence of a clearing mechanism in the futures market and also in the options market, with daily settlement of gains and losses through the margin requirement, makes these instruments the least risky.

Country risk or political risk

A large number of projects are undertaken by project promoters from different countries. Hefferman (1986) defines country risk as the risk associated with publicly guaranteed loans or loans made directly to a foreign government. This definition of country risk is narrow, as it includes only obligations, that are either a direct obligation of a government or in the nature of contingent obligation. This is also known as *sovereign risk*. However, a project in a foreign country may face a large number of other types of political risks, which may be either from the political acts of the governments of the host country or other country specific conditions. Nagy (1979) gives a much broader definition of country risk as:

> ... the exposure to a loss in cross-border lending caused by events that are, at least to some extent, under the control of the government of the borrowing country.

This definition, in fact, differentiates between the commercial risk and the country risk. Commercial risk is one that is under the direct control of the private enterprise or individual. Under country risk he includes factors that are directly or indirectly concerned with the foreign government.

Based on the above broad definition by Nagy (1979), one may identify some of the important country risk factors where the government may:

- expropriate project facilities
- withhold, delay or cancel licences required by the project for carrying out its activities
- deny an appropriate forum to resolve contractual disputes
- limit the ability of the project to convert and remit the local currency into foreign currency
- enact laws that may adversely affect the operations of the project.

In addition, the government has an influence on risks associated with war or civil disturbance that may interfere with the operations of the project. This is because it is the prime responsibility of the government to provide security from internal disturbance and war.

Accounting risk and economic risk

Before we discuss the management of financial risks it is important to understand the concepts of accounting risk and economic risk. *Accounting risk* is the risk that can be measured from an entity's financial accounts. Information about transactional cash flows, the location and denomination of assets and liabilities, and the maturity structure of the balance sheet enable an objective assessment to be made of the magnitude of risks faced by the company. Galitz (1995) explains accounting risk as:

> ... largely a backward looking concept, looking at how cash flows, assets and liabilities have been affected by risk in the past, or how they may be affected by changes that take place right now.

Economic risk goes much further, and is concerned with the broader impact of risk on an entity's entire operations. Economic risk is often concerned with second- and third-order effects, as the impact of risk ripples through the economic system.

Accounting risk is therefore relatively easy to quantify. For example, in the case of an interest rate increase, any floating rate liability will incur greater cost and any floating rate asset will earn more returns. Economic risk is, however, much wider. In case of an increase in interest rate the suppliers may demand early payments and the purchasers may demand extended credit. This may worsen a company's cash flow, leading to more borrowing and at higher interest cost. A high interest rate may slow down economic activity, leading to less demand for the company's goods. A high interest rate may lead to weakening of the currency of the country. If the company sources its inputs from a different country then there would be a higher import cost. All these direct and indirect effects can have an impact on the profitability of a company.

Summary

It is recommended that a risk management programme should be integrated within a company's overall business and financial strategy. Risk management should not be approached in an *ad hoc* manner or delegated to employees who are unfamiliar or uninvolved in formulating a company's overall strategy. It is, therefore, important that both borrowers and the lenders should adopt a risk management programme.

Bibliography

Galitz, L. *Financial Engineering: Tools and Techniques to Manage Financial Risk*. Pitman, London, 1995.

Hefferman, S. *Sovereign Risk Analysis*. Unwin Hymen, London, 1986.

Merna, T. Financial risk in the procurement of capital and infrastructure projects. *International Journal of Project & Business Risk Management*, **2**(3) (1998).

Merna, A. and Smith, N. J. Privately financed infrastructure for the 21st century. *Proceedings of the Institution of Civil Engineers*, **132**(4) (1999).

Nagy, P. J. *Country Risk: How to Assess, Quantify and Monitor It*. Euromoney, London, 1979.

CHAPTER THIRTEEN

Mechanism for risk management and its application to risks in private finance initiative projects

Introduction

The risk management process is a continuous cycle of:

- risk identification
- risk analysis
- risk response or allocation.

Risk and uncertainty are distinguished as follows:

- a decision is said to be subject to *risk* when there is a range of possible outcomes and when known probabilities can be attached to the outcome
- *uncertainty* exists when there is more than one possible outcome to a course of action but the probability of each outcome is not known.

A *hazard* has the potential to do harm or cause a loss, but the degree of risk from the hazard depends also on the circumstances. In many hazardous situations, risks can be reduced to acceptable levels by the quality of risk management applied.

This chapter looks at how a contract allocates risks between the parties to it. It identifies areas of particular concern to those parties involved in private finance initiative (PFI) projects and discusses how they may be dealt with. It examines how the payment mechanism is used in the contract to allocate risk, and how the payment mechanism itself is formulated. Included later in this chapter is a discussion on how conventional forms of contract are suited for use in the PFI.

Classification of risks

Merna and Smith (1996) broadly classify the risks associated with a project financed under project finance techniques under two categories: elemental risks and global risks.

Elemental risks

Elemental risks are the risks that are within the control of one or other of the project participants, and appropriate steps can be taken to mitigate them. These can be commercial risks of the project such as the project risk, political risk, the risks associated with the debtor (special project vehicle (SPV)), risks associated with the sponsors, and risk associated with the local government or sovereign risk.

Project risk

Project risks are numerous but they can be broadly classified into four categories: completion risk, operation and maintenance risk, input and output risk and financing risk.

Completion risks:

- the risk of delay in completion of the project due to contractors' delays
- the risk that project completion will involve cost overrun
- the risk that the project fails to meet the performance specification on completion.

Operation and maintenance risks:

- the risk that the project is unable to run at the desired efficiency due to deficiencies in equipment, personnel and maintenance
- the risk that the cost of operation and maintenance of the project turns out to be more expensive than projected
- the risk of project operation being delayed due to legal issues such as environmental liabilities
- the risk of fire and other casualties.

Risk of inputs and outputs:

- the risk of an inadequate, subquality and inconsistent supply of raw material and other utilities
- the risk of breach of contract by suppliers of raw materials and purchasers of the output
- the risk of an increase in the price of inputs as compared to the estimated price
- the risk of inadequate demand for the output of the project.

Risks related to financing of the project:

- the risk related to an increase in the servicing cost of money raised for the project
- the risk of exchange rate fluctuation
- the risk of inadequate funds in the event of cost escalation.

Political risks

Political risks are those risks either associated with the exercise of the sovereign powers of the host country government or due to certain country-specific situations. They include:

- the risk that the government will expropriate the project facilities
- the risk that the government may withhold or delay the granting of a licence or other administrative and statutory approvals for the construction of the project to start or for its operation
- the risk that the government will breach its contractual obligation to the investors and deny them forum to resolve their disputes
- the risk that the government may impose foreign exchange restrictions on the project to use foreign funds or to repatriate profits or to service foreign debt and equity
- the risk that the government may enact new laws which may adversely affect the operations of the project.

Borrower's credit risk

As compared to the project risk, which is associated with the project itself, there are risks associated with the creditworthiness of the SPV created for the implementation of the project. There is risk of default by the SPV on its liabilities due to bankruptcy or general deterioration in the overall financial condition of the company.

Sponsor's credit risk

In project financing the stakes of the sponsors are limited to the resources invested in the project by the sponsor company. The market credit rating of the SPV is different from the credit rating of the sponsor company. The involvement of known and renowned sponsors, however, gives strength to the project company, and any change in the credit rating of the sponsor company is likely to affect the credit rating of the project company, more so if the change is adverse. This can affect the cost of funds being raised by the company.

Sovereign risk

In many projects, the host country government provides guarantees or counter guarantees for the capital raised by the project company. The credit rating of the government is known as the *sovereign credit rating*. Any change in the sovereign credit rating may also affect the viability of the project.

Global risks

Global risks are the risks that are normally outside the project package and that are generally not controllable by the project participants. They are also called *force majeure risks*. They include floods, earthquakes and other natural disasters, including war and civil disturbance. In cases where the project is entirely sponsored by private participants and the involvement of the local government is minimal, the political risk becomes global risk because it goes outside of the control of the project participants. On the other hand, when the host government also participates in the project as a co-sponsor, some of the political risks, such as expropriation, become controllable and are no longer viewed as global risks.

Risk management

Risk management can be defined as any set of actions taken by individuals or corporations in an effort to alter the risk arising from their business. Risk management deals with both insurable and uninsurable risks, and is an approach that involves a formal orderly process for systematically identifying, analysing and responding to risk events throughout the life of a project in order to obtain the optimum or acceptable degree of risk elimination or control.

Risk management is an essential part of the project and business planning cycle, which:

- requires acceptance that uncertainty exists
- generates a structured response to risk in terms of alternative plans, solutions and contingencies
- is a thinking process requiring imagination and ingenuity
- generates a realistic attitude in project staff by preparing them for risk events rather than being taken by surprise when they arrive.

At its most fundamental level, risk management involves identifying risks, predicting how probable they are and how serious they might become, deciding what to do about them and, finally, implementing these decisions.

Risk identification techniques

Risk identification techniques fall into three categories:

- intuitive (e.g. brainstorming)
- inductive, i.e. 'What if?' techniques (e.g. hazard and operability studies, failure modes and effects, criticality analysis)

- deductive, i.e. 'So how?' techniques (e.g. accident investigation and fault trees, based on hindsight).

Using this categorisation as a general guide, a brief overview is given below of some of the more common methods employed. Note that most methods will involve a combination of intuition, induction and deduction, and most circumstances will warrant a combination of methods.

Brainstorming
Originated in the 1950s as a business management tool, this technique involves the collective generation of ideas by a group comprising key project and business personnel and others in an environment free of criticism. The underlying principles are that:

- group thinking is more productive than individual thinking
- the avoidance of criticism aids the production of ideas.

Group members are encouraged to 'free-wheel' in their thought association and build on each other's ideas. The most promising ideas generated are then selected, developed and verified. The classic brainstorming technique requires modification in order to apply it to risk identification, as follows:

- The objective needs to be changed from solving a problem to identifying risks. Participants must therefore have an understanding of risk management.
- The project or business requires subdivision into manageable portions to facilitate thought generation (e.g. through the use of the 'work breakdown structure' or specific areas of investment).

Limitations of this technique include its dependence on the group composition, while its simplicity and speed make it a popular approach to risk identification.

The Delphi technique
The Delphi technique was developed at the RAND Corporation for technological forecasting. It involves obtaining group consensus by the following process:

- respondents are requested to give their opinion on the risks pertaining to a project
- a chairperson collates the responses and issues a summary of the findings to the respondents, requesting that they revise their opinion in light of the group's collective opinion

- these steps are then repeated until either consensus is reached or the chairperson feels that no benefit will result from further repetition.

The respondents are isolated from one another in order to avoid personal conflicts, and interact only with the chairperson. The medium through which the process takes place tends to be either the postal service or electronic interactive media.

Benefits of the Delphi method include that participants may be remote from where the risk management process is being undertaken and are free from group pressures such as conformance, personality characteristics, compatibility and peer pressure.

Interviews

Where informational requirements are more detailed than a group can provide, or where a group work is impractical, interviews provide a means of soliciting information from individuals.

Checklists

Checklists are derived from the risks encountered in previous projects and provide a convenient means for the project manager to rapidly identify possible risks. They take the form of either a series of questions, or a list of topics to be considered. Organisations may generate checklists for themselves or make use of standard checklists available for their particular industry or sector.

Checklists are an important vehicle through which an organisation can record and capitalise on its past experience.

Risk registers

A risk register is a document or database that records each risk pertaining to a project, particular investment or asset. As an identification tool, risk registers from previous similar projects may be used in much the same way as checklists.

The risk register enables the data collected during the risk identification phase to be captured and saved, for review and as a data container for transfer of the information into the risk software of choice. There are a number of prerequisite data items necessary within the risk register:

- Project title: this should briefly describe the project.
- Project ID: this allows identification of specific projects where multiple projects are being developed.
- Activity ID.
- Risk ID.
- Activity acronym.

- Risk acronym.
- Names of the team leader and the individual team members: this information is necessary should any further future investigation be needed or should queries arise with regard to the original risk assessment.
- Activities: a list of activity descriptions, preferably in order of sequence. The register may be used for network or spreadsheet models.
- Precedence: this is important for network-based risk software packages. It identifies the linkage between the activities from start to finish.
- Most likely: estimated by the expert for the activities, this is the value used in the risk software package around which the optimistic and pessimistic values operate. This is commonly referred to a as the *three point estimate*.

Summary

The methods described above are merely an indication of those available for carrying out risk identification and are by no means an exhaustive list. In most cases a combination of methods will be employed. Furthermore, these techniques are not limited to use in risk identification, but may also be used as tools for risk analysis and risk response.

Risk analysis

Qualitative risk analysis

Qualitative risk analysis consists of compiling a list of the main risk sources (after risk identification) and a description of their likely consequences. Risks are treated independently of each other. The probability of occurrence of the identified risks is not quantified. The country risk analysis, for example, is a qualitative risk analysis in the project identification phase, which identifies and analyses political, economic, legal and social risks, without quantifying these risks.

Quantitative risk analysis

The two most widely used techniques for quantitative risk analysis are sensitivity analysis and probability analysis. These are both performed in a computer simulation programme, used in the appraisal phase of a project.

Sensitivity analysis

Sensitivity analysis and deterministic analysis checks what change in the value of a dependent variable occurs if the value of one or more variables that determine the dependent variable changes. Economic parameters such as the net present value (NPV), internal rate of return (IRR), cash pay-back period and cash balance are typical dependent variables in economic analysis. The sensitivity analysis should be done for all risks that may have a considerable impact on the project, in order to quantify the impact of those risks on the dependent variables, such as the economic parameters in economic analysis.

It is important for the project manager to pay particular attention to the variables (risks) a change in which affects the economic parameters most, not only in the appraisal stage, but also for monitoring in the implementation phase of the project. With experience, the number of variables to consider can be reduced, as the key risks with a strong impact on a project's economic parameters will become evident during the life cycle of the project. The sensitivity analysis can be used to identify the variables that need to be considered for the performance of a probability analysis.

The main limitation of the sensitivity analysis is that no indication of the likely probability of occurrence of changes in key variables is given. Another shortcoming is the fact that each variable is considered independently. In reality, management normally wants to know the result of the combined effect of changes of two or more variables because, in practice, some combination of the variables considered independently during sensitivity analysis will occur.

The above-mentioned limitations are overcome by *probability analysis*, or *stochastic analysis*, which can specify the probability (frequency) distribution for each risk and for the project. It can also consider situations where a number of variables can be varied at the same time.

The allocation of probabilities of occurrence to each risk requires the definition of ranges for each risk (often referred to as *confidence limits*). The definition of these ranges is normally subjective (especially if no historical data are available), and therefore project members who are responsible for the original estimates should be involved in this exercise.

A random sampling approach, the *Monte Carlo simulation technique*, is used. It runs the analysis a number of times and generates random numbers for each variable. The basic steps required for this technique are as follows:

- The *probability distribution* is determined for each variable that affects the economic parameter. The types of distribution available are as follows:

- *triangular distribution* reflects the fact that one is tempted to assign to a value close to the extreme of a range a lower probability than to a value close to the best estimate; this probability varies linearly from the value of the best estimate to the extreme value of the range, which makes random generation very easy
- in *stepped triangular distribution* the quantification of subjective probability judgements is based on preference ranking; it can be drawn up by the appraiser and gives the appraiser the freedom to choose whatever intervals and to divide them into as many subintervals as deemed appropriate
- *uniform distribution,* or *rectangular distribution,* is used where judgement is vague and the appraiser is not able to differentiate between any two values within the range of the variable; the use of this distribution should therefore be avoided.

Triangular distribution is widely used for ease of use.

- The *range of variation* for each variable is assessed. The range for each variable is an indication of the uncertainty of the project management's opinion concerning the degree of uncertainty about the original estimate. The larger the range (and the more flat the frequency and cumulative frequency curves) the greater is the uncertainty.
- A *value* is selected for each variable within its specified range. The frequency with which any value is selected corresponds to its probability in the distribution.
- An *analysis* is run using the value in random combinations selected for each one of the variables. With each generation of a new value for each variable a new combination is obtained. This is repeated a number of times (*iterated*) to obtain the probability distribution (*frequency distribution*) for the economic parameter.

The outcome of a project (presented as economic parameters such as IRR and NPV) is the calculated combination of these values selected for each variable. For each economic parameter a new simulation must be performed.

The greater the number of iterations the greater is the accuracy of the simulation. A number of iterations of about 200 is considered to be sufficient for accurate results. After this exercise the management should discuss risk mitigation measures and model the changes to the economic parameters generated by risk mitigation measures. The range for each variable where risk mitigation occurs can be changed to reflect the effect of risk mitigation measures on these variables. The outcome of the project (NPV and IRR) should normally improve after this reappraisal.

This exercise forces the project manager not only to think about the risks, their impacts and probabilities of occurrence, but also to think about risk mitigation measures, their implementation and effects, and thus helps the project manager to set realistic targets.

Risk response

The risk response process

Once a clear understanding of the threats and opportunities facing a project or business has been established, the final phase of risk management, the risk response process, commences. The information developed by the identification and analysis phases (What are the risks? How do they impact on project success? To what extent?) is now employed in formulating project strategy (What should be done about the risks?). Risk response is often referred to as *risk control*, which involves handling risks in a manner that achieves project and business goals efficiently and effectively.

The objective of risk treatment strategies is two-fold:

- to reduce the potential impact
- to increase control over the risks.

There are two alternative approaches to achieving this objective:

- *risk control* – the adoption of measures aimed at avoiding or reducing the probability and/or severity of losses occurring
- *risk finance* – making provisions to finance losses that occur.

The two-fold objective proposed, however, is only true of the threats or downside risks associated with a project or business, since a reduction in the potential impact of an opportunity or upside risk would obviously not be in the project's interests. However, the notion that control over risk is desirable holds in either case. The terminology used to describe the risk response process tends to imply a concentration on risks associated with loss, but the same tools and techniques may just as well be employed in the exploitation of opportunities.

The first stage of the risk response process is concerned with the development of alternative strategies to deal with the risks; then, the particular strategy or strategies most appropriate to the circumstances must be selected.

Risk avoidance

Risk avoidance involves the removal of a particular threat. This may be either by eliminating the source of the risk within a project or by avoiding projects or business entities that have exposure to the risk.

For example, a contractor, wishing to avoid the potential liability losses associated with asbestos may never acquire a project that involves operations with this material. The same scenario, but this time considered from the client's perspective, also lends itself as an example of eliminating a source of risk within a project if the risk is avoided through redesigning the facility so that it uses an alternative material to asbestos.

Risk reduction

Since the significance of a risk is related to both its probability of occurrence and its effect on the project outcome if it does occur, risk reduction may involve either lowering its probability or lessening its impact (or both). The severity of injuries from falling objects on a building site, for example, may be reduced through the compulsory wearing of hard hats, while the adoption of safer working practices can lessen the likelihood of objects falling.

Risk transfer

Projects may be seen as investment packages with associated risks and returns. Since a typical project or business involves numerous stakeholders, it follows that each should 'own' a proportion of the risk available in order to elicit a return. For instance, if a project involves the construction of a facility, some risks associated with that construction should be transferred from the promoter organisation to the contractor undertaking the work (e.g. that the project is completed within a specified time frame). In consideration of this risk, the contractor will expect a reward.

Contractual risk allocation will not be dealt with in detail here, but the fundamental considerations are the same for all risk transfer, regardless of the vehicle through which transfer is facilitated. Principles for the allocation of risk among the parties to a project are:

- Which party can best control the events that may lead to the risk occurring?
- Which party can best manage the risk should it occur?
- Which party should carry the risk if it cannot be controlled?
- Is the premium imposed by the transferee likely to be acceptable?
- Can the transferee sustain the consequences if the risk occurs?
- Do the secondary risks to the transferor result from transferring the risk?

An example of the time frame in a construction contract can be used to illustrate the above principles. The party with the greatest control over the completion date is the contractor and, as such, is in the best position to manage this risk. The client stands to lose revenue if the facility is not built by a certain date. Therefore, to mitigate any such loss, the client includes a liquidated damages clause in the contract, so that if construction overruns this date the contractor compensates the client for the loss. The contractor will consider this risk in its tender and we can expect that the contract price is higher than it would be in the absence of the clause (i.e. the transferee imposes a premium on accepting the risk). However, if the revenue loss is likely to be too great for the contractor to compensate for, there is little sense in transferring the risk in this way.

Insurance is a popular technique for risk transfer, in which only the potential financial consequences of a risk are transferred and not the responsibility for managing the risk. It should be noted that insurance costs money (premiums) and these should be balanced against the probability of the risk occurring.

Financial markets provide numerous instruments for risk transfer in the form of hedging. This is best illustrated by way of example: the fluctuation in the price of an input may be hedged through the purchase of futures options, so that in the event of a future price rise the (lower than current market value) options soften the effect. Consequently, however, the benefits of a price decrease are lessened by the cost of the futures options. Options, futures, futures options, swaps, caps, collars and floors are only some of the instruments available.

Risk retention

Risks may be retained intentionally or unintentionally. The latter occurs as a result of failure of either or both of the first two phases of the risk management process (i.e. risk identification and risk analysis). If a risk is not identified or if its potential consequences are underestimated, then the organisation is unlikely to consciously avoid or reduce it or to transfer it adequately.

In planned risk retention there is the complete or partial assumption of the potential impact of a risk. As suggested above, a relationship between risk and return exists. Without risk exposure, an enterprise cannot expect reward. Retained risks should ideally be those with which the core value-adding activities of the organisation are associated. These are the risks that the organisation is most able to manage. Retained risks should also be those that may be dealt with more cost-effectively by the organisation than by external entities (since risk transfer and avoidance must necessarily come at a premium). Finally, risk reduction may only be cost-effective up to a point, thereafter becoming more costly than beneficial.

At this stage of the risk management process, alternative risk response options will have been explored for the more significant risks. Either risk finance provisions or risk control measures, or both, for each risk now require consideration and implementation.

The corporate risk management policy

In the event that a corporate risk management policy is in existence, then decisions regarding which strategy or strategies to adopt to mitigate risks can be taken in accordance with it. In the absence of such a framework the policies, procedures, goals and responsibilities for risk decision-making need to be set, i.e. a corporate risk management policy must be established.

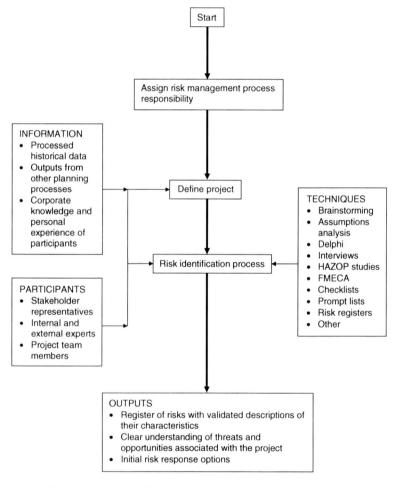

Figure 13.1. The risk identification process

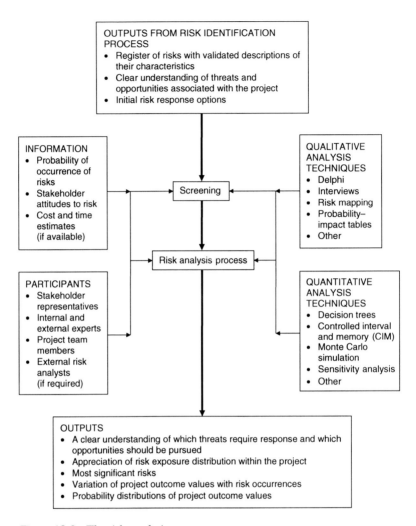

Figure 13.2. The risk analysis process

Outputs from the risk response process

Each significant risk should be considered in terms of which project party should 'own' it and which risk response options are suitable for dealing with it. The most appropriate response option or options in accordance with the corporate risk management policy, and consequently the response strategy or strategies, must then be selected.

The processes of risk identification, analysis and response, which constitute the mechanism of risk management, are summarised in Figures 13.1 to 13.3.

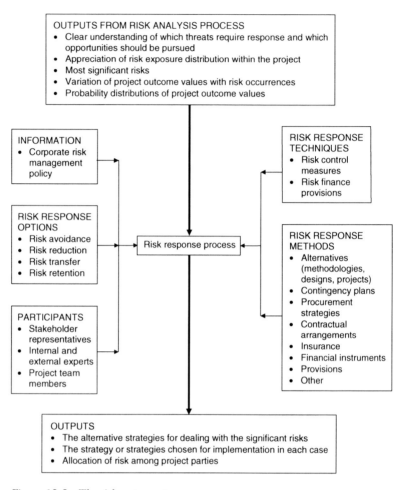

Figure 13.3. The risk response process

Risk transfer in PFI projects

The phrase that should be dominant when considering risk transfer is: *the risk should be allocated to the party who is best able to manage it.* In order for PFI to be successful, there must be risk transfer on some scale, but this does not mean that total transfer should be the objective. The optimum allocation is not where the public sector shoulders zero risk. Risk transfer and its magnitude is highly dependent on the type of PFI project being considered (e.g. joint venture, free standing, service provision).

Taking a PFI service provision contract as an example, we can identify the following as a selection of major risk headings:

- design and construction
- commissioning and operating
- demand
- residual value
- technology/obsolescence
- regulation/legislation.

These risk types are discussed below in the context of a PFI service contract.

Design and construction risks

The main implication of design and construction risks is to the cost and time element of the project. In PFI projects these risks must always be transferred to the constructor, as the contractor is always in the best position to control and manage these risks. A client can further protect himself from these risks by acquiring performance bonds and/or parent company guarantees. In addition to this, liquidated damages can be sought, but these are usually of small value in comparison to the problems caused, and in such circumstances the liquidated damages may not offer sufficient protection. A client can further protect himself by carefully structuring the payment mechanism. Under the PFI these factors play an influential and important part of risk allocation.

Construction and design risks can be further broken down into:

- co-ordination of design and construction
- site conditions
- industrial relations
- construction difficulties
- fitness for purpose of the design
- functionality
- adaptability
- constructability
- maintainability.

Commissioning and operating risks

Commissioning and operating risks are a fundamental feature of a PFI. The acceptance of these risks by the operating SPV is the incentive for them to operate effectively and more efficiently than the previous public sector management.

Operational risks include:

- availability of materials
- reliability of equipment
- unpredicted demand

- productivity levels
- irregular revenue patterns
- interruption of business.

Demand risks

Demand risks are possibly one of the most contentious areas of risk allocation. They are heavily dependent on what kind of project is under consideration. This is best illustrated by an example.

HM Prison Service is a monopoly purchaser of prisoner accommodation in the UK. In addition, HM Prison Service has total control over where prisoners are sent and for what proportion of their sentences. Hence the operating company has no control over whether or not any of the available space is to be occupied. It would seem unreasonable for concessionaires to run the risk of demand when it is entirely beyond their control. In such a situation there would have to be an equitable payment mechanism, which has an element of remuneration that is independent of occupancy levels.

Residual value risk

The client would wish the operator to control the residual value risk. By doing this the operator is guaranteeing that the facility will be maintained to a sufficient standard, as it is in his best interests to do so. Without this situation the operator may be tempted to neglect maintenance, as it would provide little or no financial benefit to him, especially towards the end of the concession period.

The BOOT dilemma

If the client wants to adopt a build–own–operate–transfer (BOOT) strategy it is difficult for him to hand over residual risks and take ownership at the end of the concession period. One way around this is by adopting an approach similar to that used on the NIRS2 project. The National Insurance Records Service is a PFI contract let to Anderson Consulting to design, finance, install and operate a National Insurance computer system for the Inland Revenue.

Here the operator was given a shortened contract, with the opportunity to re-bid for the right to run the system under the contract. If a new operator took over the existing system the original operator would receive a transfer payment. If the new operator provided a new system, then no payment would take place. This is designed to provide an incentive to the original operator to fully maintain the system, as it is in his own best interest.

At the end of the second contract the client has the option to take ownership or place it out to tender again. Under these circumstances it is hoped that the client obtains the flexibility of ownership and opens the contract up to negotiation, while ensuring that the system is properly managed and maintained. A similar approach was taken for the London Underground Northern Line, with trains having a life of 35 years, but the contract let for 20 years, allowing for competition to occur mid-way through the life of the facilities.

Technology obsolescence risk

Technology obsolescence risk is not always an issue in PFI projects, as the objective is to supply a fully functional service or facility. As long as the service is not affected by the technology in use and continues to provide the specified service there is little problem. However, in projects that are technology intensive, it would be wise to operate under a shorter concession period to allow competition and technical advances to enhance the service. This was done in the NIRS2 project outlined in the previous section. If state of the art technology is to be employed, then it is useful to benchmark the service against an external system. This would ensure that the system would not become obsolete in comparison to other systems. Payments could be linked to this benchmark, encouraging the operator to continually upgrade and improve the system. This is particularly relevant to information technology (IT) and information system (IS) projects, which are being proposed.

For many projects, particularly financially free-standing ones such as road and bridge projects, the technology obsolescence risk will not constitute an issue. There may be some degree of upgrading required, such as the transition from shadow to direct tolling on the emergence of a reliable electronic tolling system, but for the most part even an obsolete project will satisfy the requirements demanded of it.

Regulation and legislation risks

Consortia seek to protect themselves in the contract with the use of clauses that reflect direct legislative changes in their specific industry. For example, if a toll bridge or tunnel is being operated and the government decides to introduce massive taxes on private transportation, this would potentially damage the operator's cash flow. The public authority is likely to resist such a contractual position, claiming that legislative change risks are all part of everyday business. Such risks are negotiated during the pre-completion stage of the agreement.

The Private Finance Panel (PFP) published a book on standard contract clauses, which aimed to reduce time spent in the drafting of PFI contracts (see Merna and Owen, 1998). Clauses were included on the basis of lessons learnt during the early stages of the PFI and calls by all parties for a standardisation in contracts to cover PFI. The following list gives the topics of the clauses contained in the PFP guide:

- entire agreement
- law and jurisdiction
- waiver
- independent contractor
- contractor's responsibility and knowledge
- assignment and subcontracting
- provision of information
- confidentiality
- Official Secrets Act
- publicity
- security
- contractor's personnel
- personal data
- payment
- value added tax
- payment of subcontractors
- *force majeure*
- termination on insolvency
- consequential arrangements on insolvency
- corrupt gifts and payments of commission
- racial discrimination
- health and safety.

Form of contract in PFI projects

A large factor in PFI projects is the number of interfaces between contracts on a typical contract. To overcome this management problem it was suggested that the New Engineering Contract (NEC) group of contracts could be used to combat the problem. The NEC was developed to provide a system that is flexible enough to cope with large and complex projects, but can still encourage good project management of the many interfaces involved. It is possible that a PFI contract could be added to this suite of contracts or that the NEC system could be made to fit each PFI project.

Payment mechanism

The payment mechanism is the formula used to decide what the concessionaire receives in return for providing a service to the end user. The fine details of the mechanism are the source of debate and intensive negotiations. It is usual for a single (unitary) payment to comprise three factors:

- provision or availability
- demand or usage
- quality.

It is generally felt that this method offers reasonable risk allocation to both parties, but also provides incentive to the concessionaire to maintain service levels through a well understood and monitored quality criteria.

Provision or availability

By paying a concessionaire for the amount of capacity he is providing, the public body accepts a limited amount of demand risk. This seems equitable, as in most cases the public body is a sole purchaser and has control over demand aspects. It also means that a concessionaire will continue to make available maximum capacity in the knowledge that he will receive (limited) benefit from doing so.

Demand or usage

The usage aspect of the payment reflects the actual quantity of service used by the public body. A value will be allocated to each 'unit' of service (however it may be defined) used over a specific time scale and a value reached. This payment reflects actual demand and presents the greatest risk to the concessionaire under the payment mechanism system. If the public body declines to use the facility or service, then no revenue will be accrued by the concessionaire under this payment element.

Quality

In the terms of the PFI agreement, strict quality guidelines will need to be set out for the concessionaire to achieve. This stream of the unitary payment reflects the concessionaire's performance against quality criteria and seeks to transfer the risk of poor performance solely on the concessionaire. Some public sector organisations were worried that in order to win PFI contracts, SPVs would tender unfeasibly low prices, which would eventually translate to reduced service quality provision. By linking a comprehensive quality threshold to the payment mechanism it is hoped that poor performance can be limited.

Interaction of the payment mechanism

Even though the principle of the three streams contributing to a single payment seems a straightforward and simple matter, it is setting the weighting and relative importance of each factor that presents difficulties. The blending of three independent criteria

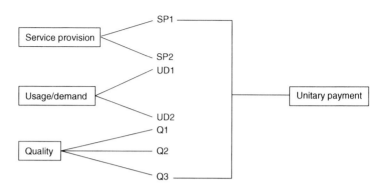

Figure 13.4. Payment mechanism

to produce single meaningful revenue for the concessionaire is complex and time consuming. Close attention must be paid to the examination of priorities and the consequences of a multitude of different scenarios.

Conflicting interests are viewed from each part of the agreement. The operators would like to see the emphasis placed on service provision, as they are unable to control or influence demand, while users would prefer a payment mechanism based on volume or usage linked to service quality. It is in the negotiating phase prior to contract award that these factors are decided in order to produce equitable risk allocation. The simplified and generalised diagram in Figure 13.4 indicates how weighting could be applied to various criteria, all contributing towards the final unitary payment.

Penalty clauses

A penalty system can be introduced to provide further incentives to the consortium to fulfil their objectives in respect of the service the client demands. Such a system could be used in conjunction with the existing quality aspect of the payment mechanism, but potentially may involve less administration and audit resources. The penalty system could be used to address small but persistent and unsatisfactory performance characteristics that a quality audit may not pick up. By allocation of weighted points to specific quality violations, a fine levied and/or payment can be withheld once a predetermined threshold has been reached.

The subject of specific violations would, obviously, vary from project to project and across sectors, but the following example is presented to illustrate the concept. In a hospital the payment mechanism would be looking at achieving priority service facilities:

- number of available beds
- waiting times for appointments
- quality of patient care
- nurse/patient and doctors/patient ratios
- cleanliness
- hygiene
- catering
- ward size, and other service parameters.

The penalty system would look to factors not directly associated with the core operations of a hospital. These factors could include the ease of navigation through the building using signposts, the external appearance of the building, the number of car parking spaces for visitors, the provision for visitors' refreshment and the number of seats made available in casualty. All these factors, and many more, are influential in the smooth operation of a hospital, but it would be complicated to adjust payments monthly to take them all into account. Using penalty points and fines it is still possible to monitor such factors but only punish the operator when a prescribed number of offences have been recorded.

Progress has been made in this area with the National Health Service (Private Finance) Bill, which addresses the concerns of over 70 NHS projects, with a value of nearly £¾ billion. The Bill enables:

> ... externally financed development agreements where the relevant finance is provided otherwise by the trust.

It is now hoped that these projects will go ahead and pave the way for success in the massive potential of the PFI market within the health service. Further legislation is also likely to be introduced to allow similar problems to be removed from local authority projects, although this will be dealt with in a separate bill.

The OPEC risk

Founded over 40 years ago by Iraq, Iran, Kuwait, Saudi Arabia and Venezuela, the Organisation of Petroleum Exporting Countries (OPEC), which controls 40% of the world's oil production, first sent shock waves through the world economy in 1973 by announcing a 70% rise in oil prices and by cutting production. The effects were immediate, resulting in petrol shortages in many parts of the world. In the UK, motorists queued for petrol and the government issued petrol ration books, although formal rationing was never implemented. In December, faced with a simultaneous miners' strike, the UK government put the country on a 3-day week, ended television transmission at 10.30 p.m. and ordered shops to turn off display lights at night.

With OPEC's production cuts pushing oil prices even higher, by January 1974 they had quadrupled, the UK's slide into recession began. As inflation passed the 25% mark in 1975 and the economy sank deeper in the mire, the stock market lost two-thirds of its value, the banking system came close to collapse and many projects were either shelved or stopped. A second oil shock in 1979 when oil prices increased again in the wake of the overthrow of the Shah of Iran appeared to confirm OPEC's power. A third sharp rise in oil prices also occurred in 1990 as a result of the Gulf War.

However, when North Sea oil came on stream, so did other areas where production would have been uneconomic in the pre-1973 cheap oil era. This diminished OPEC's control on oil pricing. Three years ago oil prices plunged to $10 a barrel and appeared to be heading even lower. OPEC, which now had 11 members, seemed unable to exercise control to halt the slide. Three things came to OPEC's rescue. Economic recovery in Asia meant a rise in demand for oil. Low prices had resulted in widespread cutbacks throughout the oil industry and, most surprising of all, OPEC agreed to cuts in production quotas and stuck to them.

The west's economies now use about half the amount of oil to produce a given amount of gross domestic product than they did in the 1970s. Even with this reduction and harnessing of technology gains and energy from alternative sources, industry and projects still rely heavily on oil. The risk associated with fluctuation in the oil price cannot be dismissed at any time when assessing the commercial and economic viability of an investment. At the time of editing this book, oil prices have fallen from US $30 to below US $20 per barrel, even after the events of 11 September 2001.

Summary

This chapter has demonstrated the importance behind understanding project risks and the need to allocate them to the parties who are best able to manage them. The major headings of risks, which have emerged as important in negotiating for PFI projects, has been explained. This chapter has also:

- described briefly the UK government's attempt to introduce standard clauses for adoption into PFI contracts
- discussed the possibility of using a modified NEC form of contract
- described how the payment mechanism works and interacts to produce a single unitary payment to the concessionaire

- discussed how a penalty system can work alongside the payment mechanism to monitor performance
- discussed the recent progress to counter the ultra-vires problems.

Bibliography

Merna, T. and Smith, N. J. *Projects Procured by Privately Financed Concession Contracts*, vol. 1, 2nd edition. Asia Law and Practice, Hong Kong, 1996.

Merna, T. and Owen, G. *Understanding the Private Finance Initiative*. Asia Law and Practice, Hong Kong, 1998.

CHAPTER FOURTEEN

Insurance and bonding

Introduction

Insurance and bonding are important, especially when considering the financing of all infrastructure projects. Their availability to pay for serious loss or damage can be crucial to the procurement of a successful project. This chapter identifies and describes a number of insurance and bonding methods often employed in engineering projects. In this chapter the terms *bonds* and *guarantees* are used extensively and are deemed to have the same meaning.

Insurance

Under traditional public works contracts the client, usually a government body, places a contractual requirement on the contractors to take out and maintain certain insurances; normally contractor's all risks third party liability and employer's liability and workers compensation. Beyond checking that the contractor has provided the client with certification of insurance, the client would traditionally take no further significant part in the insurance arrangements for the contract.

In most construction projects insurance policies are required. The type of insurance policies that provide the cover required by the Institution of Civil Engineers (ICE) conditions are material damage policies and liability policies.

The only *material damage policy* required by the ICE conditions is the *contractor's all risks policy*, which protects the insured's property and that for which he is responsible (i.e. the works). The material damage policy indemnifies the insured contractor against physical or material loss or damage from whatsoever cause to the contract works or materials whilst on contract site(s) and in use and where such loss or damage arises out of the performance of the contract and/or during the period of maintenance.

The two main *liability policies* that cover the insured's legal liability are the employer's liability policy and the public liability policy. The *employer's liability policy* is compulsory, and not only ensures a fund for valid claims by employees, but also prevents insurers from repudiating liability under their employer's liability policies because of a breach of certain terms and conditions. The operative clause covers the liability of an employer to his employees, who are persons under a contract of service or apprenticeship with the employer, for bodily injury or disease arising out of and in the course of employment. The *public liability policy* provides an indemnity against bodily injury and property damage claims by the public (other than bodily injury claims by employees). Where the employer's liability and public liability policies of a contractor are with different insurers it is necessary to ensure that the scope of risk excluded from the public liability policy matches exactly the risk included in the employer's liability policy, otherwise a gap in cover occurs.

While it is not practical to examine each of these individual policies in detail, the process used by insurers in evaluating one of the more significant classes is examined to gain a better understanding of how policy terms and conditions are determined. The following factors represent some of the main underwriting considerations that insurers would examine before offering a quotation for cover during the construction phase of a project, relative to construction all risks liability and third party liability insurance.

Fire
The possibility of damage caused by fire is usually one of the first risks that insurers assess. Many contracts are performed in locations far from proper fire-fighting facilities, and there may be an inadequate supply of extinguishers and buckets of sand and water on site. Warning notices may be unheeded, and the training of fire crews inadequate.

Explosion
The possibility of explosion is increased by the use or storage of chemicals or by the application or presence of heat in confined places.

Subsidence, collapse and land slip
Weakening of the support of a building or structure is normally caused by disturbance of ground conditions on adjacent land. It often causes subsidence, collapse and other severe damage, and figures prominently in connection with tunnelling work and locations with geological faults.

Earthquake
Insurers have ready access to plentiful information on the severity of earthquakes in various parts of the world. The more significant

include the Rocky–Andes mountain chain in the Americas, North Africa, Middle East, Japan, The Philippines and New Zealand.

Rainstorm and flood
The most serious damage occurs usually when a rainstorm causes rivers and other stretches of water to break their banks, or causes water to drain from a large natural catchment area and flood lower lying areas. Much damage can be caused by flash floods in areas where the normal rainfall is very low. The risk factors to be considered include: the level of the contract site in relation to that of the surrounding area; its closeness to the sea, a river or other stretch of water; the maximum level or rainfall to be expected in the area in 24 hours; and the expected seasonal variations in the water table on the site. Flood warning or flood control systems may be in operation in some areas, but their efficacy cannot be taken for granted.

Windstorm
The major windstorm areas are the Arabian Sea and Bay of Bengal, Australasia, China and the North Pacific Zone, the south-west Indian Ocean, the south-west Pacific, the West Indies, the Gulf of Mexico and the southern USA. The design of a structure will usually provide greater safety when the structure is completed than during construction.

Frost
This peril is likely to cause damage only if allowed to do so by a failure to take adequate precautions.

Seasonal perils in general
It is also necessary to consider the seasonal perils separately. An 18-month construction period may include one or more well-defined rainstorm or windstorm seasons. The progress chart will show which construction work coincides with adverse weather and temperature seasons. A careful note should be kept of any such calculations, so that they are available should the contractor request an extension of the construction period, running into an additional hazardous season and meriting more than a *pro rata* additional premium. In a number of cases temperature derivatives have provided options based on degree days paid off when the temperature reached a certain aggregation of abnormally hot or cold days over a given time period. This derivative was seen as suitable for certain projects where temperature data were easy to model and price using standard actuarial techniques.

230

Overbalancing, fall, impact, crushing and friction

Impact may be caused by, for example, the overbalancing of a crane, the movement or position of its jib or load, or the dropping of its load. Impact is also caused by the negligent use of other mechanical plant, particularly motor vehicles, on the site. The risk of damage due to impact by aircraft or other aerial devices or articles dropped therefrom is not important, except in the vicinity of airports.

Infidelity and theft

Theft is facilitated by poor security systems and by collusion between the insured's employees on the site and visiting subcontractors' employees, particularly motor vehicles on the site. The risks can be reduced by having security patrols, watchmen and locked steel cages to contain loose fittings. Articles most likely to be stolen include non-ferrous metals, expensive fittings and plate glass.

Riot, civil commotion, and malicious and terrorist acts

These perils depend on political, economic and sociological factors. The most reliable guide is the recent experience in the locality of the site. In some countries riots tend to occur in the heat of the summer.

Breakdown or derangement of machinery

Breakdown is one of the main perils to be considered during the testing and commissioning phase of process projects.

Trends in global insurance

During the last 10 years or so the global insurance market has been going through one of the most difficult and unpredictable states within living memory. In part this is the consequence of intense competition in both the traditional and emerging marketplaces since the mid-1980s. In the last 10 years downward pressure on pricing has been cushioned by high interest rates. Now that this safety net has been removed, the true effect of the poor trading results of many insurers and re-insurers has been starkly revealed.

The difficult trading conditions experienced have been exacerbated by a well-documented series of major catastrophes, for example:

- 1987, storm and floods in Europe
- 1988, Piper Alpha platform
- 1989, Hurricane Hugo and *Exxon Valdez*

- 1990, winter storms Vivian and Daria in western and northern Europe
- 1991, Typhoon Mirielle
- 1992, Hurricane Andrew.

In more recent times there have been:

- earthquakes in Japan, Turkey and Taiwan
- rail disasters in the UK and Germany
- tunnel fires in the Alps and the Channel Tunnel
- hurricanes on the USA eastern seaboard
- winter storms in Lothar in western Europe
- north ridge earthquake in the USA
- heavy flooding in the UK and also in western and central Europe
- Typhoon Bart in Japan.

Special note

At the time of editing this chapter, two airliners crashed into the twin towers of the World Trade Centre in New York, and a third airliner crashed into the Pentagon in Washington, DC, on 11 September 2001. The insurance bill for this incident (an apparent terrorist attack directed on the USA) is likely to run into billions of dollars. The massive death toll, the loss of four airliners and the collapse of the twin towers and other buildings in Manhattan will undoubtedly begin the largest ever claim in the 300-year history of the insurance industry. Insurers are likely to face claims for damage to the World Trade Centre and the Pentagon and for business interruption during the rebuilding period. The latest press reports suggest that the final cost for this historic attack, considered an act of terrorism, could be as high as US $100 billion, the major part of the claim being the cost of personal accident and life insurance.

Insurers and re-insurers around the world are likely to suffer massive losses, especially Lloyd's of London following its disastrous pay outs of recent years. The total bill will dwarf the US $20 billion, and still rising, which insurers paid out as a result of Hurricane Andrew, the most costly insurance claim up to now. More than US $5 billion was wiped off the value of Britain's leading insurance companies and the value of shares fell on all international stock markets immediately after the attack. The price of gold bullion, however, rose from US $273 to US $288 per ounce, and the price of crude oil increased from US $27.4 to over US $30 per barrel.

The effect of these occurrences, both individually, but more significantly cumulatively, has been extremely damaging to the capital strength of the insurance market. To this must be added other specific problems, such as the losses incurred by the UK Composite, Life & Non-Life Insurers as a result of mortgage indemnity claims in the UK, major terrorist-related losses in the UK and the USA and the effect of the collapse in the French commercial property market on the balance sheets of the major French insurers.

The consequences of these trading difficulties have created the following difficulties in obtaining adequate insurance protection for projects:

- The market, due to restrictive reinsurance constraints, is restricting cover for the following risks: gas turbines of all types, loss or damage caused by defective design, workmanship and/or materials, and loss or damage due to natural perils in specific geographical locations (e.g. parts of South America, Japan, the Philippines).
- A significant reduction in capacity for delay in start-up insurance.
- Higher levels of self-insurance are constantly being requested, specifically for natural perils and design risks.
- The cost of procuring insurance protection is increasing dramatically, and it is becoming essential that accurate insurance prices are included within the project bid, in order that financial projections can be properly measured by financiers.

Another insurance policy adopted by consultants and designers involved in projects is the *professional indemnity policy*. A professional indemnity policy protects the insured against his legal liability to pay damages to third parties who have sustained some injury, loss or damage due to the negligence of the insured or his staff in the professional conduct of the business. Normally, liability is based on breach of professional duty, which is the failure to exercise an ordinary degree of care and skill in the profession concerned. There is only a limited insurance market available and there is no standard policy.

Non-traditional insurance techniques have been developed to cover credit enhancement and fitness of purpose. Cat Bonds, for example respond not to cash losses, but to precisely measurable phenomena, such as Richter scale or wind speed, translated to some cash payout.

Bonding

Many client organisations, in both the public and the private sector, stipulate that a bond is provided by a contractor. Many main contractor organisations also insist that nominated subcontractors and other subcontractors provide them security by provision of a bond in a similar manner to the obligation of the main contractor to the employer. The stability of the party providing the bond must be beyond question, and for this reason many employers prefer the provision of the bond to be made by the insurance market.

When banks are asked to participate in the debt financing of a project they will look at the percentage of the guaranteed and insured loans. Obviously, a higher rate of guaranteed loans will attract more banks into the financing, to the advantage of the project sponsors. Typical guarantees may include the following:

- tender guarantees
- performance guarantees
- advance payment guarantees
- retention bonds
- completion guarantees
- maintenance bonds.

Tender guarantees assure the contract issuer that the tendering party will accept the tender if selected. Premiums for these guarantees are typically 0.5–1.0% of the tender value.

In *performance guarantees*, a performance bond is usually provided at contract award to safeguard against poor performance by the contractor. The value of work guaranteed is normally 10% of the contract sum, with premiums typically being 1–2% per annum of 100% of the contract sum. Normally, the value does not reduce, but performance bonds should have an expiry date (not necessarily a calendar date – it can be linked to an event so that time slippage is automatically taken into account). If the value of the contract increases or the duration of the contract extends, the bond may need to be amended accordingly. In some cases parties may specify that the performance guarantee covers a maintenance period provided for in the contract. A performance guarantee will not in itself ensure that projects are carried out efficiently and to time, but it will be one of a number of commercial pressures on the contractor to perform well.

An *advance payment guarantee* allows the issuer to regain any advance payments made to the contractor in the event of him not fulfilling the contract terms. Where the advance payment reduces with time, as for example stage payments made against goods and/or services delivered under the contract, the full value of the advance payment is guaranteed, with premiums of up to 1–2% per annum of that sum.

Retention bonds are provided so that contractors (and their subcontractors) may be paid without the client deducting retention money. As work is completed, the contractor is paid fully under the terms of the contract. The traditional retention system is to withhold a percentage from payments made during the course of the contract in order to accumulate a fund that is available to the clients if the contractor fails to rectify defects in accordance with the contract (usually 5% of the value of the contractor's work up to certified completion, reducing to 2.5% up to final acceptance, the remainder being paid at the end of the maintenance period). It should be noted that retention bonds are intended to cover two situations: first, where the bond is given upon practical or substantial completion against the release of retention money held at that date; and, secondly, where the bond is given at the beginning of the contract period, in which case it would generally have a cumulative value up to practical or substantial completion, at which point the value would be reduced by 50%. The use of retention bonds transfers financing cost from the contractor to the client (who is required to pay in full earlier) and will pass cash flow benefits to the contractor. Their use will only result in a lower cost to the client if contractors are prepared to reduce their tender prices accordingly. The option to offer a retention bond should be included in the tender documents at tender stage.

Completion guarantees ensure that the contractor will complete the works, or specified areas of work, by a specified date. These dates may be specified for tests, normally indicated on the construction programme, before the specified completion date of the contract. In the UK retention monies or the performance bonds normally cover completion guarantees.

Maintenance bonds protect the employer and are an alternative to retention monies during the maintenance period. They guarantee that once the contract works have been completed the contractor will carry out his obligations under the maintenance period or defect liability period of the contract. In the UK, however, maintenance guarantees are usually linked to the performance guarantee or through retention monies.

Table 14.1 illustrates the typical costs, duration and cover of the bonds described in this section. The variance in premiums and cover often depends on the form of contract and the perceived risk in the project.

In general, bonds fall into two broad categories: on-demand or conditional. In the case of *on-demand bonds* the beneficiary can require payment of the bond merely by asserting that payment is due and without proof of default or condition, other than procedural. A *conditional bond*, on the other hand, is only callable when the beneficiary can 'prove' default on the part of the contractor or supplier.

Table 14.1. *The typical cost, duration and cover of bonds*

	Type of bond					
	Bid bond	Advance payment guarantee	Retention monies guarantee	Performance bond	Completion bond	Maintenance bond
Typical bond value (% of contract value)	Fixed sum (usually 5% of budget price)	10%	0.5–2.5%	10%	2.5%	2.5%
Typical duration of coverage (months)	Tender period + 3 months	Variable, as dependent on work programme	12 months after substantial completion or variable from contract start	Contract period + 12 months	Practical completion + 12 months	Practical completion + liability period
Typical bond cost (% of bond value)	0.5–1%	1–2% per annum	1–1.5% per annum	1–2% per annum	0.5–1% per annum	0.5–1% per annum

It should be noted that the premiums paid by contractors are normally determined on the basis of their past performance and financial standing for a guarantee facility, rather than being a standard amount. On-demand bonds often require a greater premium than do conditional bonds.

Issuers of bonds

In the UK the use of bonds has been fairly limited. They have in the main been confined to public sector work. Only recently has bonding been arranged for private sector employers and between main contractors and their subcontractors.

Banks and insurance companies

Bonds are issued by banks and insurance companies. There are advantages and disadvantages to both methods. Banks regard bonds as an extension of their line of credit to the contractor. They usually know the contractor well – much better than an insurance company would, for example. Banks can, therefore, issue bonds speedily, but the bond when issued will affect the line of credit, perhaps prejudicing the advance of money in the future to purchase plant and machinery. Banks usually pay more promptly as they are less willing to argue the merits of any claim than is an insurance company, and do not normally make the same investigations as an insurance company would.

Banks are the most usual source of guarantees in the UK, often through their own insurance departments. Banks see bonds as a normal part of the service that they can offer to their clients, and are currently trying to standardise the terms of bonds with regard to obligations, coverage, call and commitment, security and format. Often the bond is for a greater value than the net worth of the contractor. The bank's understanding of the contractor's business is, therefore, very important. In most cases on-demand bonds are provided by banks who may regard them as open credit notes and may require provisions to be made from borrowing facilities against contingent liabilities.

Before an insurance company will issue a bond it makes careful enquiries into the contractor's financial stability and his ability to perform the contract. These enquiries made by insurance companies are helpful to the employer, since he knows that if an insurance company has issued a bond it is satisfied with the risk. Conversely, if the insurance company refuses to issue a bond then the employer should ask whether he has chosen the right contractor. Insurance companies are a main source of guarantees in Australia, New Zealand and Scandinavia.

Surety companies

Surety companies are virtually the only source of guarantees in the USA and Canada. A US surety company is quite prepared to accept responsibility and arrange for a contract to be completed by another contractor in the event of the default of the original contractor. Because of this situation the surety company may well intervene in the negotiations for a contract, and also during the running of the contract. The surety company would also investigate very thoroughly whether the employer had fulfilled all his obligations under the contract, in establishing his own responsibility under the guarantee. Guarantees offered by surety companies differ substantially from the indemnity concept, in that the emphasis is on the surety arranging completion of the project and not just on paying the amount of the bond following default.

Where the contractor is part of a larger company or group of companies, it is often a requirement of the employer for the parent company to guarantee the performance of the contractor. Such a guarantee is free of cost to the client, but may give less certainty of redress than a bond because it is not supplied by an independent third party. However, whilst accepting less independence, parent company guarantees for the right contract can be more advantageous than bonds. Rather than receiving a fixed amount in compensation, the parent company is obliged to complete the contract. Costs for completion are borne by the parent company, which may be significantly more than the compensation provided for in a bond. In addition, further recompense can be sought for time delays in completion through the normal clauses incorporated in the contract.

The conditions of a parent company guarantee will usually give the parent company the opportunity to remedy default within a period of notice before the guarantee is called. The liability can take several forms, including a financial guarantee of completion of the project itself or the employment of another contractor to complete the project.

Where problems arise under the contract, this form of guarantee should discourage the parent company from putting the contractor into liquidation solely to avoid losses in completing the project or in paying damages for late or non-completion. Because the financial strength of the parent company may be linked to that of the contractor, a parent company guarantee will be acceptable only if the parent company is financially strong and its financial resources are largely independent of those of the contractor.

The guarantee should be for the entire performance of the contract without limitation. 0–0.5% is a common premium calculated on the sum guaranteed per annum.

Summary

All projects require an element of insurance and bonding. The number of insurance policies or bonds employed in a project will often depend on its complexity and the demands made by the clients. The cost of insurance and bonding depends on the type of project, commercial market conditions and on a contractor's financial standing. The cost of bonding is borne by the client, and the more bonds required by a client the greater the project cost. It should be noted that excessive insurance and bonding do not turn a bad project into a good one.

Bibliography

Eaglestone, F. N. and Smyth, C. *Insurance under the ICE Contract.* George Godwin, 1990.

HM Treasury. *Bonds and Guarantees*, CUP No. 48, Central Unit on Procurement, 1994.

Madge, P. *Civil Engineering and Bonding.* Thomas Telford, London, 1987.

Merna, T. and Smith, N. J. (eds). *Insurance, Projects Procured by Privately Financed Concession Contracts*, 2nd edition. Asia Law and Practice, Hong Kong, 1996.

CHAPTER FIFTEEN

Case study of a toll bridge project

Background information

A developing country is proposing to construct a bridge across a river in order to ease the city's traffic problems. In particular, there are major traffic problems between the road linking the airport to the commercial and financial district of the city. The country has enjoyed an annual growth in gross domestic product (GDP) of 8–9% over the last 8 years and is categorised by the World Bank as an upper middle income country. As a result of the improving economy, sales of motor vehicles have increased.

The road connecting the airport to the city centre has suffered heavy damage due to high usage by an increasing number of vehicles, and the only way to ease the traffic problem is to construct another link (bridge) between the two parts of the city. However, the fiscal resources of the government are being stretched to the limit by more urgent needs for housing, hospitals, schools and other infrastructure services to cater for the increasing population. After consideration, the government proposed that the bridge project be carried out on a build-own-operate-transfer (BOOT) basis. The project will be procured by inviting the private sector to undertake the design, construction finance, operation and maintenance of the bridge.

Traffic projection surveys conducted by the local traffic department show that the current daily traffic volume is approximately 90 000 vehicles, and this number is expected to increase further in the next 10–15 years.

The proposed concession will be awarded through competitive tendering. Several potential promoters have been invited to evaluate and submit their tenders. A particular promoter proposes the following bid. The data given in the following section are the financial estimates of the project.

Computer aided simulation for project appraisal and review (CASPAR)

CASPAR is a software package developed by the Centre for Research in the Management of Projects at UMIST. The CASPAR program utilises the network approach to project modelling. The network

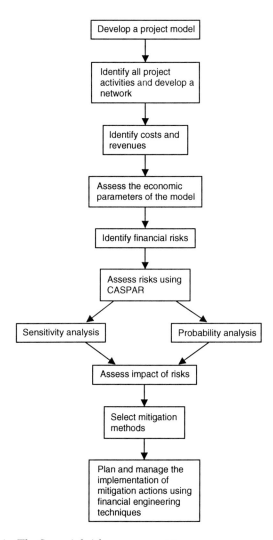

Figure 15.1. The financial risk management process

modelling technique has been used to model and prepare realistic outcomes of many projects procured by privately financed concession contracts. The software is designed to model the interaction of time, resources, cost and revenue throughout the entire life of the project. The program enables a model to be developed which incorporates the identified risks and simulates their impact on the project cashflow. It allows models to be generated showing sensitivity to specific risks and also models the probability of them occurring. Figure 15.1 shows a typical financial risk management process for a project.

Financial estimates for the project

Project: proposed construction of six-lane toll bridge
Concession period: 26 years
Construction period: 6 years (including feasibility studies, prelimi-
nary design, estimate, traffic survey and purchase of land)
Total cost of construction: US $300 million
Estimated maintenance cost of the bridge: US $3 million per year
Total maintenance cost of the bridge: US $3 million × 20 years =
US $60 million
Expected revenue generated: US $2268 million over the concession
period
Exclusivity: 20 years
Taxation: 35%; tax may be waived for the first 5 years.

The promoter has set a minimum acceptable rate of return (MARR)
of 12% for this project. In order to simplify the models, the costs and
revenues in future years are assumed to be the same as the current
cost and revenue, thus avoiding discounting and the effects of inflation. Inflation would be countered through a tariff variation formula. The interest rate (for the debt), the coupon rate (for the
bonds) and the rate of return (on the equity) for investors are
assumed to be as shown in Table 15.1.

Finance package

The promoter develops five financial models to determine which is
the best financial package for the project. In this project, the three
main financing instruments (debt, equity, bonds) are used. In addition, the promoter considers an interest-free grant from the host
government as one of the options in financing the project. The government has agreed to provide grants up to a maximum of 5% of the
total construction costs, payable at the end of the concession period.
The five financial models are described below:

1. This model (Figure 15.2) uses conventional financing in the
    form of the debt/equity ratio. A debt/equity ratio of 80:20 has
    been adopted for the purpose of this case study.

Table 15.1. Interest rate, coupon rate and rate of return

Financial instrument	Interest rate (per annum)
Debt	8%
Bonds	7% pre-tax
Equity	9% pre-tax

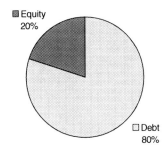

Figure 15.2. Financial model 1

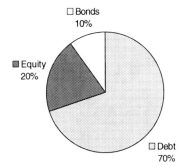

Figure 15.3. Financial model 2

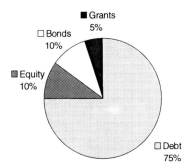

Figure 15.4. Financial model 3

2. In this model (Figure 15.3) bonds are considered as another form of financing, together with debt and equity.
3. In this model (Figure 15.4) it is assumed that grants are available for the project, in the form of aid from the government. The amount of grant is assumed to be at a maximum rate of 5% of the total construction costs.

4. This model (Figure 15.5) is similar to model 2, except that the amounts of each financial instrument considered are different. In this model, 60% of the project cost comes from debt, while the remaining 40% of the project cost come from bonds (20%) and equity (20%) in equal proportions.

5. In this model (Figure 15.6) bonds are not considered. Instead a considerable amount of debt is used.

In this case study we considered five different models in an attempt to create several different financing scenarios. All five models have similar financing instruments, but have different proportions of debt, bond, equity and grant financing. The five financial models are used to calculate the economic parameters of the project, which provide a basis for comparison between various simulations carried out. Later, financial risks are identified and input into the computer program (CASPAR) to determine the effects of those risks on the five models.

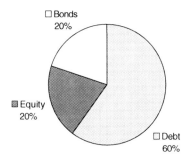

Figure 15.5. Financial model 4

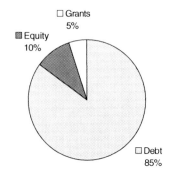

Figure 15.6. Financial model 5

Table 15.2. Economic parameters of the five financial models

Economic parameter	Financial model				
	Model 1	Model 2	Model 3	Model 4	Model 5
NPV (US $ million)	874.41	869.01	893.25	844.53	908.13
IRR (%)	14.60	14.85	14.90	14.94	14.74
Payback (years)	11.20	10.99	11.08	10.79	11.28

IRR, internal rate of return; NPV, net present value

Limitations of the financial models

There are limitations to the five financial models. They do not take into account the following:

- the cost of raising finance through equity issue and bond issue
- the cost of using financial engineering techniques to manage financial risks.

Deterministic analysis

The economic parameters relating to the five models were analysed using CASPAR. The results of the analysis are summarised in Table 15.2. The economic parameters determined are the net present values (NPV), the internal rate of return (IRR) and the payback period.

As the table shows, model 5 has the highest NPV at the end of the concession period, while model 4 has the highest IRR and the shortest payback period. However, the deterministic results do not provide conclusive evidence as to which is the most suitable financial package for the project, as the financial risks associated with the project have not been taken into account and have not been analysed.

In this project, the main criterion for the promoter is the IRR, rather than the NPV and the payback period, as this can be compared with the promoter's minimum acceptable rate of return of 12%.

Financial risk analysis

According to Table 15.2, models 3 and 4 appear to be the most suitable package for the toll bridge project, as they have the highest IRR and the highest NPV, respectively. However, there are many factors that control the financial viability of the project. Structuring the financial package of the project requires a detailed analysis of the financial risks that are identified at the appraisal stage. The effects of the financial risks may be greater in model 3 and model 4 than in other models, which in turn may affect the IRR and the NPV of the project.

In this project, the following financial risks are assumed:

- change in construction cost
- change in interest rate
- change in market demand
- change in tax rate
- change in equity cost
- change in exchange rate
- change in bond payments.

Two types of risk analysis techniques are used in the case study: sensitivity analysis and probability analysis.

Sensitivity analysis

Sensitivity analysis is a deterministic modelling technique that is used to test the effect of changes in the value of each independent variable (risk) on the project. It involves a relatively simple diagram where one parameter value is changed at a time while holding the other parameter values constant.

Sensitivity analysis provides answers to a wide range of 'What if?' situations. For example, what happens to the internal rate of return if the demand for the project decreases? What happens to the cost of financing if the currency of that country is devalued?

The focus of this type of analysis is to draw attention to the most serious risk that may affect the project, thus narrowing down the main variables to be considered. However, the disadvantage of sensitivity analysis is that only one risk is considered at a time and does not actually evaluate risk or the effects that one risk may have on another.

Probability analysis

Probability analysis overcomes the limitation of sensitivity analysis by specifying a probability distribution for each risk, and then considering the effects of the risks on the economic parameters of the project in combination. The risks are identified and a range of values for each risk is estimated. A suitable probability distribution is chosen for each risk. A value within the range is selected for each risk. This value should be within the probability distribution. There are four common probability distributions: uniform, beta, rectangular and triangular distributions. In this project, the triangular distribution is adopted in the five financial models. Triangular

Table 15.3. Risk specification table

Risk	Distribution	Lower limit	Upper limit	Parameter
Change in construction cost (CCM)	Triangular	−10	+20	Construction cost
Change in interest rate (CIR)	Triangular	−10	+30	Debt finance
Change in market demand (MTD)	Triangular	−50	+50	Revenues
Change in tax rate (CTR)	Triangular	0	+20	Tax payments
Change in equity cost (CEC)	Triangular	−100	+50	Equity finance
Change in exchange rate (CER)	Triangular	−20	+20	Cost of finance
Change in bond payments (CBP)	Triangular	−100	+50	Bond payments

distribution is applied to a risk variable if it can be predicted with some confidence, and if the estimate used in the model is most likely, or slightly optimistic. In this project the range of values applied to the risk variables are predicted with some confidence that they are most likely to happen.

The outcome of the project is calculated by combining the values selected for each risk. A computer is used to calculate the combined impact of the input values and to generate a probability distribution of the possible outcomes of the project. The computer software used in this case study is CASPAR, which uses the Monte Carlo technique. The analysis carried out on the five financial models is based on 1000 iterations, to ensure that there is no sampling bias. The output of the analysis gives a range or estimates for the outcome of the project, ranging from the most pessimistic to the most optimistic outcome.

In this project the promoter is more concerned about the probability of getting the MARR of 12% from the financial models after the models are analysed. The IRR is the most applicable economic parameter to consider for a long-term project such as this toll bridge project (Merna and Smith, 1996). The model that has the highest probability of getting an IRR of 12% within the specified frequency range will be the most suitable model for this project. In this project, a frequency range of 15–85% is adopted. The perceived risks, the range of variations and the relevant parameters are illustrated in the risk specification table (Table 15.3).

The upper and lower limits are different for each risk variable. This is due to the nature of the risk and the level of exposure that the project is subjected to. For example, the lower limit for tax rate is zero, as it is envisaged that the host government will not lower the tax for this project. However, the host government may increase the tax.

Construction projects are likely to cost more or less than the initial estimates due to delays and cost overruns. Thus −10 to +20 is proposed as a reasonable lower and upper limit.

Interest rates are more likely to increase rather than decrease, giving the −10 to +30 lower and upper limit. The +30 upper limit is based on the assumption that the economy of the country could overheat, resulting in the government increasing interest rates to combat inflation.

Market demand is considered to be the most important factor in the success of the project as the contract is a market-led infrastructure project. It is assumed that the number of motorists using the bridge would vary, thus the −50 and +50 lower and upper limit. Although traffic projection surveys conducted by the local traffic department show an increasing number of vehicles, there could be a possibility that motorists do not want to use the toll bridge.

The lower and upper limits for the exchange rate are based on the assumption that the local currency is quite strong and stable and would not depreciate or appreciate against the US dollar by more than 20%.

For the cost of equity and bond payments, there is a risk that the project may not generate enough revenue to pay any dividends and coupon payments to investors after debt service, and thus the −100 lower limit. If the project generates more revenue, the dividends paid to the investors would be higher. However, coupon payments will be fixed. A weaker than expected revenue generation might lead to a fall in share prices and will inhibit the project's ability to raise additional funds should the need arise.

Results of the sensitivity analysis

In a sensitivity analysis diagram the gradient of the line represents the sensitivity of the project to the risk variables. If the line of a particular risk variable is close to the vertical axis, the project is very sensitive to the variable. If the line of the risk variable is close to the horizontal axis, the project is least sensitive to the variable. Also, the project is very sensitive to the risk variable if a percentage change in the variable produces a greater percentage change in the IRR of the project.

Due to the number of variables and financial models, combining all the variable curves in a single diagram is quite difficult.

Therefore the analysis result of each model is shown separately. The results of the sensitivity analyses of the five financial models are shown in Figures 15.7 to Figure 15.11.

Market demand

From Figures 15.7 to 15.11 it is clear that the project is most sensitive to the change in market demand for the services, which is the number of motorists using the toll bridge. Any changes in market demand will have a very significant financial impact on the project, in terms of the IRR.

In financial model 1 a drop of 50% in market demand in terms of revenues leads to a drop of about 51% in the IRR of the project. In financial model 2 a decrease of 50% in market demand reduces the IRR of the project by 49%. In financial model 3 a 50% decrease in market demand reduces the IRR by 47%, while a 50% decrease in market demand reduces the IRR of financial models 4 and 5 by 51% and 47%, respectively.

Exchange rate

The project is financed mostly in foreign currency, thus exposing the project to foreign exchange risk. It is important that exchange

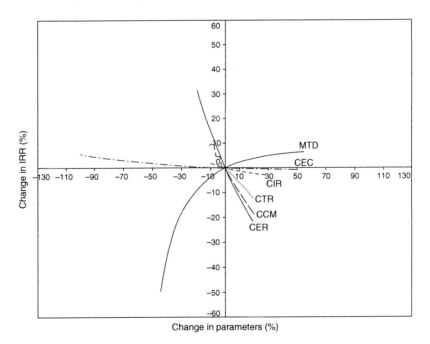

Figure 15.7. The results of the sensitivity analysis for model 1

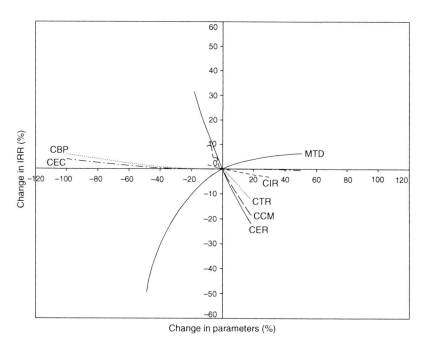

Figure 15.8. The results of the sensitivity analysis for model 2

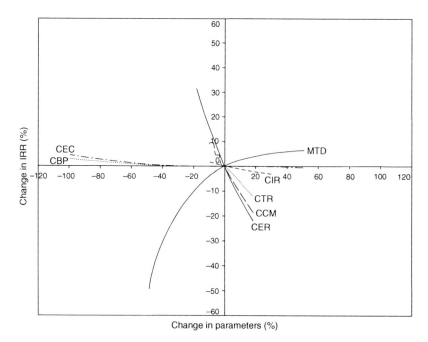

Figure 15.9. The results of the sensitivity analysis for model 3

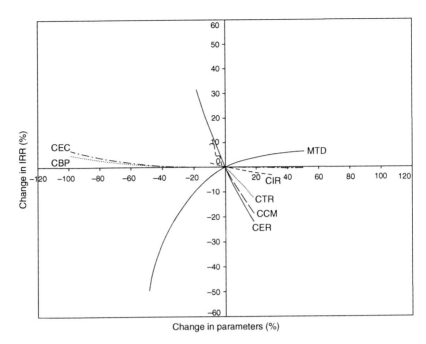

Figure 15.10. The results of the sensitivity analysis for model 4

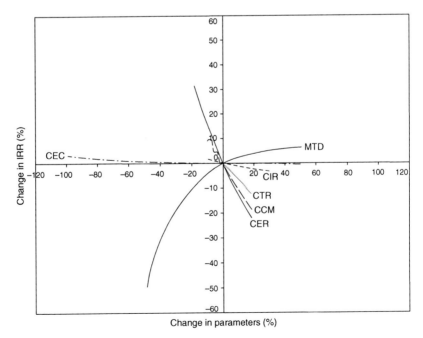

Figure 15.11. The results of the sensitivity analysis for model 5

rate fluctuations should always be considered when appraising the financial viability of the project. Any weakening in the local currency will result in an increase in principal and interest repayments, the dividend payment and the cost of construction, since more money is needed to buy foreign currency.

In model 1 a 20% drop in the value of the currency leads to a drop of about 22% in the IRR of the project. Model 2 shows a drop of 22% in the IRR if the value of the currency drops by 20%. A 20% drop in the value of the currency leads to a drop of 21% in the IRR for model 3, of 22% in for model 4 and of 21% for model 5.

The risk of currency exchange has greatly affected the viability of this project. Thus the promoter should pay considerable attention to managing this risk

Cost of construction
The next risk variable to which the project is most sensitive is the change in the cost of construction. Due to the large financial investment required to undertake this project, it is important that construction is completed within budget and without any cost overruns. In financial model 1 a 20% increase in construction cost leads to a drop of about 19% in the IRR of the project. Model 2 shows a drop of 19% in the IRR, while models 3, 4 and 5 show a similar drop of 19% in the IRR. This would invariably affect the viability of the project.

Tax payments
The IRR of the project is sensitive to tax payments applicable to the project revenue. In every country there is a corporation tax on the earned revenue. A 35% tax is levied on the revenue stream of this project. Therefore, there is an effect on the IRR of the project. Model 1 shows that a 20% increase in tax rate leads to a drop of about 13% in the IRR of the project, while models 2–5 show that a 20% increase in tax rate reduces the IRR by a similar percentage.

Interest rate
The project is sensitive to changes in interest rates. An increase in interest rate will result in a higher debt repayment, which will in turn affect the IRR. In model 1 a 30% increase in interest rate will reduce the IRR of the project by 3.7%. Model 2 shows a reduction of 3.3% in the IRR, while models 3, 4 and 5 show a reduction of 3.3%, 2.8% and 3.7%, respectively. From these results it can be concluded that the higher the proportion of debt financing, the higher the risk that the project is exposed to interest rate changes.

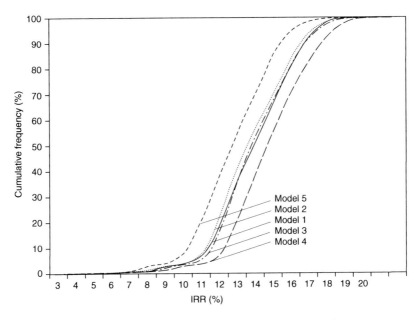

Figure 15.12. The results of the probability analysis for models 1–5

Cost of capital

The effects of the costs of equity and bond payments do not have quite such a significant impact on the IRR of the project as do the other risk variables. In models 2, 3 and 4, a 50% increase in the bond component reduces the IRR of the project by about 0.3%, 0.2% and 0.3%, respectively. Models 1 and 5 do not have bond components in their package. Similarly, a 50% increase in the equity component results in a similar reduction in the IRR for all models.

The reason for all the five financial models having similar results is due to the same financial risks being applied to all the models. The range of values applied to each risk variable is constant for all models, and therefore similar results are obtained.

Results of the probability analysis

The results of the probability analysis are represented in the form of cumulative frequency diagrams. The frequency diagrams of the five financial models are represented in a single diagram (Figure 15.12) to compare their range of values for the IRR which the project may achieve and to determine which model has the highest probability of achieving the promoter's initial MARR of 12%.

Table 15.4. The approximate range of the IRR for models 1–5

Financial model	IRR range (%)	% of achieving a MARR of 12%	Mean IRR (%)
1	10.0–14.8	55	13.1
2	10.9–15.6	64	13.2
3	9.1–14.9	53	13.3
4	11.0–15.8	68	13.4
5	11.0–15.5	68	13.3

The cumulative frequency curve shown in Figure 15.12 illustrates the outcomes of the five financial models in terms of the IRR over the frequency range 15–85%. The approximate range of the IRR for each of the five models is given in Table 15.4. According to the results given in the table, financial models 2, 4 and 5 have the highest percentages of achieving the promoter's MARR of 12%. Models 4 and 5 have the highest IRR range of the five models, while model 4 has the highest mean IRR.

It is suggested that the reason for getting similar results with all five models is because the same financial instruments were applied to the models. Although the proportion of financing is different, the proportions are not sufficiently different to generate different results. However, financial model 4 would be the one most likely to be selected for this project because of the higher mean IRR and the higher IRR range obtained with this model.

Summary

In this case study five financial models were proposed for the privately financed toll bridge. Different financial instruments were used in these financial packages, the three main ones being debt, bond and equity. All five financial models use different proportions of financing utilising these financial instruments.

The toll bridge project is a long-term, market-led infrastructure project. Therefore, the financial risks associated with the project are large. The risk that the project is most sensitive to is the change in market demand. Clearly, the viability of this project depends on the level of accuracy of the market research undertaken by the promoter. This risk, together with all the other financial risks, reduces the commercial viability of the project. The risks are applied to all the financial models to assess the effects that the risks have on the models.

To determine the best financial package for the project, an investigation of the risks was carried out by using a computer software program (CASPAR) to assess the economic effects of the risks on the project. CASPAR was used to perform sensitivity and probability analyses on all the five models. An initial MARR of 12% was determined. The financial model that has the highest probability of achieving the MARR of 12% is deemed to be the most suitable model for the project.

Bibliography

Merna, T. and Smith, N. J. *Projects Procured by Privately Financed Concession Contracts*, vol. 2. Asia Law & Practice, Hong Kong, 1996.

CHAPTER SIXTEEN

Case study on managing project financial risks utilising financial engineering techniques

Introduction

In the case study considered in Chapter 15, five financial models were analysed using CASPAR to determine the best finance package for the project. Model 4 was selected as a result of the analysis. To ensure that the project is carried out smoothly, the financial risks discussed in that case study must be carefully managed. It is essential to manage the risks at the beginning rather than wait for an event to happen and then rely on all parties to resolve them. In this second case study several financial engineering techniques and other solutions are discussed and used to manage the financial risks discussed in Chapter 15.

A result of the financial risk management will be that the ranges of the risks will be reduced and new results will be tabulated after performing sensitivity and probability analysis.

Change in construction cost

In this project the construction cost includes the risk of delays and cost overruns. Generally, the best way to manage this risk is for the promoter to use fixed price, turnkey contracts, with provisions for liquidated damages if the contractors fail to complete the bridge on time. This helps to control the project cost. This will also provide security for the lenders, as a turnkey contract seeks to complete a project on time. The promoter can include contingency sums in the original cost estimates in case of cost overruns, and maintain standby credit facilities (standby loans). Standby credit facilities will incur additional costs to the project.

This risk is usually borne by the promoter, as he can control the construction process. The promoter, in turn, would pass the risk to the main contractors, subcontractors and suppliers. Lenders would also ensure that 'step-in-clauses' are included in the loan agreement

Table 16.1. The borrowing costs of the promoter

Payments	Receipts
LIBOR + 2% (to lenders)	LIBOR + 1.5% (from party A)
Fixed rate 7% (to party A)	

to remove a contractor not carrying out the work to time, cost and specifications.

Change in interest rate

Caps and floors are normally used to prevent interest rates from rising to a level at which the project has problems repaying the debt. For example, the promoter borrows at a floating rate of LIBOR + BP, where LIBOR is the London Interbank Offer Rate and BP is basis points. To protect against higher interest rates, the promoter also buys an 8% 5-year interest rate cap. So, if the LIBOR + BP is below 8% the promoter simply pays the prevailing interest rate. If the LIBOR + BP is higher than 8% the promoter would only pay at the cap level, which is 8%. If the promoter is confident that the interest rate will not fall below 6% he can sell a floor at an interest rate of 6% to the lender. However, if the interest rate falls below 6% the promoter would have to pay the lender the difference in interest rate.

Interest rate swap is also a useful financial engineering technique to manage interest rate risk in project finance. To prevent interest rates from rising and incurring a higher interest payment, the promoter can enter into an interest rate swap with another party. For example, the promoter can borrow at a fixed rate of 8% and a floating rate of LIBOR + 2%. Say the promoter chooses to borrow at floating rate of LIBOR + 2% and agrees to pay party A a fixed rate of 7% on a notional principal of US $180 million (75% debt). In return, party A agrees to pay the promoter LIBOR + 1.5% on the same notional principal.

Under the swap, the promoter has reduced his borrowing cost from 8% to 7.5%. The swap has also eliminated the interest rate risk. When the interest rate rises, the receipts of payment from party A offset the cost of borrowing. Table 16.1 shows the promoter's borrowing costs under an interest rate swap over the loan period.

The promoter can also use a forward rate agreement (FRA) to manage the interest risk. For example, he can fix his interest obligation by buying an FRA now for a loan to be taken out in 6 months' time. Assume that the promoter arranges an FRA with the bank and that the interest rate for 6 months ahead is 8%. In 6 months' time, if the interest rate remains at 8%, the promoter borrows at the going rate of 8%. If the interest rate rises to 8.5%, the bank pays the promoter an equivalent amount to compensate for the difference in the interest rate of 0.5% (8.5% − 8%). However, if the interest rate falls to 7.5%, the promoter would have to pay the bank the difference in the interest of 0.5% (8% − 7.5%).

Change in market demand

In this project, market demand is the biggest risk faced by the promoter and lender. The promoter faces the prospect of motorists not using the bridge, which could greatly reduce revenue generation, as the main source of revenue comes from toll receipts collected from motorists. This project is slightly different from other projects, such as power stations, as this is a market-led infrastructure project. Unlike a power project, the bridge has no off-take agreement with motorists. The project could face a serious problem of repaying the loan if the revenue is lower than expected. The lender, on the other hand, faces the risk of default by the promoter.

The risk in this project cannot be mitigated using conventional methods, as the usage of the bridge is entirely up to motorists. The host government can help reduce the promoter's exposure to the risk by providing governmental guarantees of a minimum operating income. The host government can also provide additional finance in case the promoter experiences a drop in traffic volume during the concession period. For example, the Malaysian Government guaranteed additional funds to PLUS, the promoter of the North–South Expressway, with the provision that if there is a drop in traffic volume in the first 17 years of operation the government will make up the deficit (Tiong, 1990).

Change in tax rate

The host government can consider imposing a fixed tax rate on the project to make the project more attractive to the promoter, or bring the tax rate lower than the standard corporation tax to boost the promoter's revenue income. However, it is unlikely that tax would be lowered. The host government can consider granting tax relief for the project.

Change in exchange rate

The revenue collected in this project is in local currency and repayment of loans is in foreign currency, thus exposing the project to an exchange rate risk. Exchange rate risk can be mitigated in the ways described below:

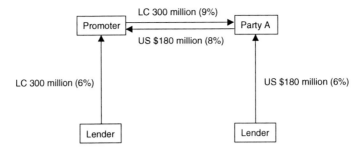

Figure 16.1. The currency swap arrangement (LC, local currency)

Box 16.1. Interest rate payments of the promoter

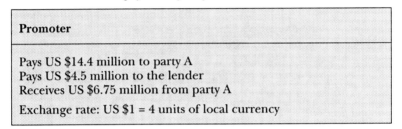

Mixing local currency and foreign currency

The exchange rate risk can be reduced by using local currency to cover costs (operating and maintenance costs, wages) so that the project does not rely excessively on foreign funds.

Using a currency swap

If there is a local currency swap market, local currency can easily be swapped with a foreign currency to remove the project's currency risks. Besides reduced borrowing costs for the promoter, a currency swap also helps to eliminate the risk of currency exchange.

For example, the US lending rate for the promoter is 8% and the local currency lending rate is 6%. To reduce its borrowing cost, as well as to eliminate the currency exchange risk, the promoter can enter into a currency swap with another party (say party A). Let us assume that party A needs to borrow 300 million of local currency. Let us also assume that the US lending rate and local currency lending rate for party A is 7% and 9%, respectively. The promoter borrows 300 million in local currency at the rate of 6%, while party A borrows US $180 million at the rate of 7%. Figure 16.1 shows the currency swap arrangement. The exchange rate between the US dollar and the local currency is US $1 = 4 units of local currency. Box 16.1 shows the interest rate payments for the promoter.

By engaging in a currency swap, the total interest rate payment is US $12.15 million. This is a rate of only 6.75% per annum, compared to 8% if the promoter has to borrow on his own. In addition, the promoter has eliminated the risk of currency exchange by borrowing in local currency.

Using forward contracts and options

Forward contracts and options are the other methods that can be used by the promoter to manage the currency risk. For example, if the promoter has to repay part of its loan in US dollars in 6 months' time, he can enter into a 6-month forward contract and fix the rate for buying US dollars today. This will protect the promoter against any movement in the local currency. However, using this method has its disadvantages. The promoter will not benefit if the US dollar weakens against the local currency, as he has to pay to compensate for the difference in exchange rate between the US dollar and the local currency.

The promoter can also consider buying an option. This will give the promoter an option to buy the option in the exchange market if the exchange rates are favourable to the promoter. For example, the promoter buys a call option to buy US dollars at a specified exchange rate within a period of time. Assume that the exchange rate is US $1 = 4 units of local currency. If the exchange rate is US $1 = 3.5 units of local currency, the promoter can exercise that option to buy US dollars at a lower rate. Similarly, the promoter can buy a put option to sell US dollars at a specified exchange rate within a period of time.

Using futures

The promoter can eliminate currency exchange risk by using futures. For example, the promoter can enter into a 30-day futures contract in which the promoter agrees to buy US dollars in 30 days' time at a specified exchange rate between the US currency and the local currency. So, if the exchange rate fluctuates sharply during the period, the promoter would not be affected by the fluctuations.

Change in the cost of equity and the bond payment

The risk of share prices falling is beyond the control of the promoter, since the share price depends on the performance of the project. However, the promoter can issue bonus shares below market prices to existing shareholders to retain the interest of these investors.

A debt–equity swap can be used to convert some of the bonds into equity. In this way a lesser coupon payment will be made, and the

Table 16.2. The risk specification table, as the mean values of the response to the identified risks

Risk	Distribution	Lower limit	Upper limit	Parameters
Change in construction cost (CCM)	Triangular	−10	+5	Construction cost
Change in interest rate (CIR)	Triangular	−10	+10	Debt finance
Change in market demand (MTD)	Triangular	−40	+40	Revenues
Change in tax rate (CTR)	Triangular	0	+20	Tax payments
Change in equity cost (CEC)	Triangular	−100	+25	Equity finance
Change in exchange rate (CER)	Triangular	−10	+10	Cost of finance
Change in bond payments (CBP)	Triangular	−100	+25	Bond payments

new equity holders have an opportunity to share profits in the future. However, the project still runs the risk of defaulting on its coupon payment and not announcing a dividend. The new risk specification table is shown in Table 16.2 as a mean of response to the identified risks.

The risk of cost overruns would be limited by using a lump sum turnkey contract and imposing liquidated damages. The upper limit is lowered from +20 to +5. The lower limit is unchanged, as the cost of construction may be reduced if the project is completed earlier than scheduled, or if the specification is reduced.

By using caps and floors and other financial engineering techniques such as interest rate swaps and FRAs, the upper limit can be reduced. The lower limit is unchanged as the interest rate might fall. In this case, the promoter will not gain if interest rates are below a certain level. However, the promoter will be certain that at the lower interest rate he will be able to meet his financial responsibilities should this lower rate be allowed.

The risk of a fall in the number of motorists is offset partly by the host government providing governmental guarantees of a minimum operating income. This risk is not fully hedged, and thus a reduction in the lower limit from −50 to −40 is suggested. The upper limit is changed to +40, due to the anticipation that the volume of traffic will increase in the future, but not as much as initially estimated.

The tax rate is unchanged, since it is assumed that the host government will not lower the tax. The possibility of tax relief is explored in the next section.

The risk of currency exchange is managed by using several methods: mixing local and foreign currency, currency swaps, forward contracts, options and futures. The lower and upper limits are set at −10 to +10, respectively.

The lower limit for the costs of equity and bond payment is unchanged, as there is still a possibility of not paying dividends and coupons. The upper limit is reduced from +50 to +25.

Sensitivity and probability analysis after financial risk management

The new results after risk response are depicted in Figures 16.2 and 16.3. The sensitivity diagram shows that the project is still most sensitive to market demand. A 40% drop in revenue reduces the internal rate of return (IRR) by 26%. There is a reduction in the ranges of the other risks, as there is more confidence in the limits of the ranges after risk management.

The probability diagram (Figure 16.3) indicates that the possibility of achieving the promoter's minimum acceptable rate of return (MARR) of 12% is about 91% after risk management, compared to 68% before risk management. This shows that the risk management has greatly increased the financial viability of the project.

Financial engineering

Although financial model 4 has been selected for the bridge project, the promoter can explore other ways to make the project even more financially viable and also to structure the financial package to make it look more attractive and appealing to the project lenders. Here, three options are explored:

- the effect of an extended concession period
- the effect of tax relief for the first 2 years
- the effect of issuing zero coupon bonds.

The effect of an extended concession period

The promoter can explore the effect of extending the concession period by 1 year to see if a higher IRR can be achieved. The results are shown in Table 16.3.

Figure 16.4 shows the probability analysis diagram when the concession period is extended. Extending the concession period has

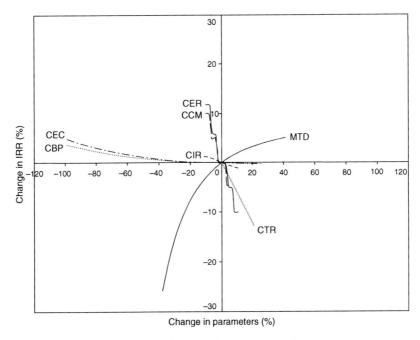

Figure 16.2. The results of the sensitivity analysis for model 4, after risk management

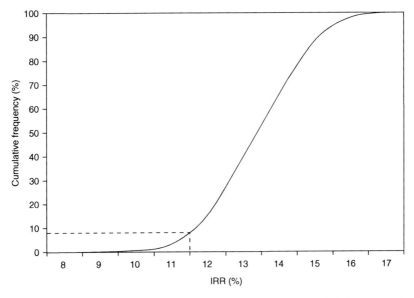

Figure 16.3. The results of the probability analysis for model 4, after risk management

increased the viability of the project. The probability of achieving an IRR of 12% is now 98%. It is even possible that an IRR of 13% can be achieved. The diagram shows that the probability of achieving an IRR of 13% is approximately 89.5%.

The effect of tax relief for the first 2 years

Significantly, the host government or the principal is often the deciding factor in the realisation of a build–operate–transfer (BOT) project. Therefore, to make the project more attractive to the promoter, the host government or principal can grant a tax relief for the project during the first 2 years of operation. This should increase the viability of the project. The effects of tax relief are shown in Table 16.4.

The effect of tax relief, often called a *tax holiday*, has greatly increased the IRR of the project. Figure 16.5 shows that by granting

Table 16.3. The effect of an extended concession period

Economic parameter	Financial model 4
NPV (US $ million)	903.0
IRR (%)	15.03
Payback period (years)	10.81

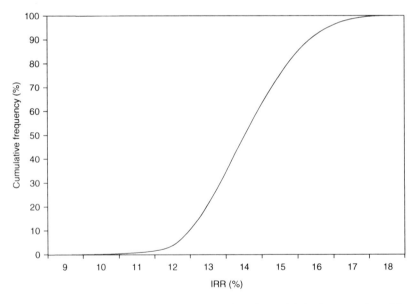

Figure 16.4. The results of the probability analysis for model 4, after extension of the concession period

Table 16.4. The effect of tax relief for the first 2 years

Economic parameter	Financial model 4
NPV (US $ million)	906.0
IRR (%)	17.66
Payback period (years)	9.47

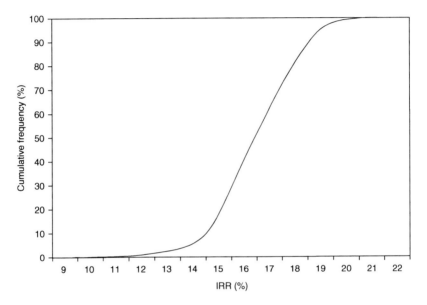

Figure 16.5. The results of the probability analysis for model 4, after applying tax relief for the first 2 years

a tax relief the promoter can realise an even higher rate of IRR. The figure shows that there is an 89% chance of achieving an IRR of 15% and a 97% chance of achieving an IRR of 14%.

The effect of issuing zero coupon bonds

Instead of issuing bonds and paying coupons every year, the promoter can issue zero coupon bonds. In zero coupon bonds, investors do not receive any interest (coupon). Income comes entirely from a bond appreciating in value. In this case study the value of the bond is assumed to increase by 7% each year, which is the same as the coupon rate on a plain vanilla bond. The effects of issuing zero coupon bonds are shown in Table 16.5.

Figure 16.6 shows the probability analysis diagram of the effect of issuing zero coupon bonds. The effect of issuing zero coupon bonds has slightly increased the possibility of attaining the MARR of the project. There is now a 96% chance of achieving an IRR of 12% and an 85% chance of achieving an IRR of 13%. Although the promoter has to pay out a considerable amount of money in redeemed bonds at the end of the bond maturity date, the effect of not paying coupons during the concession has relieved the promoter's debt burden. However, since the pay-outs for the redeemed bonds are high, the net present value (NPV) of the project is greatly reduced.

Probability analysis after risk response and financial engineering

Figure 16.7 shows the probability analysis diagram after risk response and financial engineering. It can be seen that there is now a

Table 16.5. The effect of issuing zero coupon bonds

Economic parameter	Financial model 4
NPV (US $ million)	735.0
IRR (%)	15.11
Payback period (years)	10.74

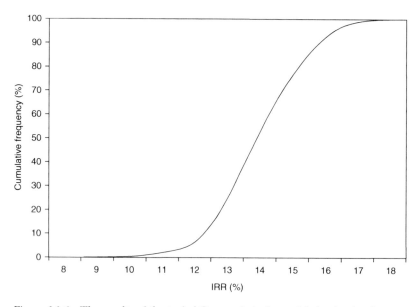

Figure 16.6. The results of the probability analysis for model 4, after issuing zero coupon bonds

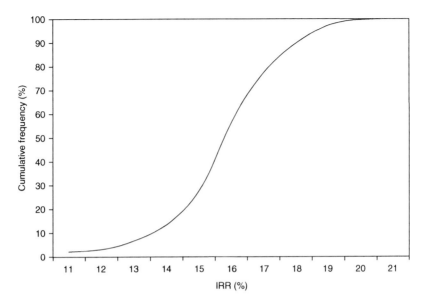

Figure 16.7. The results of the probability analysis for model 4, after risk mitigation and financial engineering

92% chance of achieving the promoter's MARR of 12%. In addition, there is also a 95% chance of achieving an IRR of 13% and a 91% chance of achieving an IRR of 14%. This shows that by mitigating financial risks in the project and by applying financial engineering to the project a higher IRR can be achieved.

It is interesting to note that the IRR is higher when only the tax relief is introduced into the project. While tax relief on the project greatly increased the IRR, the introduction of an extended concession period and the issuing of zero coupon bonds somehow lowered the IRR of the project. This could be due to the high pay-out of principal to the bondholders at the end of the concession period, which in turn affected the NPV and IRR of the project.

Summary

In the case study considered in this chapter, several financial engineering techniques were used to manage the risks in the project. Forward rate, FRA, interest rate swap, currency swap, futures, options, caps and floors were used to manage the risks. However, it should be noted that the costs of using these techniques were not included in the case study. Therefore, the promoter should weigh up the advantages and disadvantages of using these techniques.

Besides risk management, several methods were considered to increase the commercial viability of the project. The effect of extending the concession period, tax relief and issuing zero coupon bonds was to increase the commercial viability of the project and to increase the possibility of achieving an IRR of 12% or more. Tailoring the finance around the project makes the project more robust and more viable to the lenders and the promoter.

Conclusion

Due to the dynamic nature of infrastructure projects, the spectrum of risks will change considerably during each stage of the project. This is due to changes in the nature of risks as the uncertainties are decreased and different financial supports are applied to the project.

The risks associated with the market and political environment are usually large. Inflation and foreign exchange rates often combine with the cost of capital to reduce the viability of the project and limit the benefits that could be achieved by an increase in the concession period. The tax payable on revenues also has a significant effect on the IRR of the project. The early stage of the operation phase is extremely vulnerable to change and requires detailed study.

Many of the global risks associated with projects carried out in most developing countries are not quantifiable and are difficult to represent accurately in a project model. Most elemental risks, which are directly related to the project activities, are often included in the sensitivity analysis. Although it is expected that a change in government would affect the country policy in political, commercial, legal and environmental areas, these factors are also difficult to quantify and were not quantified in this project model.

Many of the global risks, including political, legal, environmental and some commercial risks, are assumed to be borne by the principal, as they are outside the control of the promoter and their effects are difficult to model. In most cases these global risks are allocated through the concession agreement between the principal and the promoter.

Appraisal of the case study presented in Chapter 15 and that presented in this chapter confirms the widely held view that host governments play a crucial role in the development of privately financed infrastructure projects, and have the potential to make projects succeed or fail.

268

Bibliography

Merna, T. and Smith, N. J. *Projects Procured by Privately Financed Concession Contracts*, vol. 1. Asia Law & Practice, Hong Kong, 1996.

Tiong, R. L. K. Comparative study of BOT projects. *Journal of Management in Engineering*, **6** (1990).

Index